CALIFORNIA NATURAL HISTORY GUIDES

FIELD GUIDE TO
PLANT GALLS OF CALIFORNIA
AND OTHER WESTERN STATES

California Natural History Guides

Phyllis M. Faber and Bruce M. Pavlik, General Editors

Field Guide to

PLANT GALLS

of California and
Other Western States

Ron Russo

Photographs and Illustrations by Ron Russo

UNIVERSITY OF CALIFORNIA PRESS
Berkeley   Los Angeles   London

To Sheri

University of California Press, one of the most distinguished university presses in the United States, enriches lives around the world by advancing scholarship in the humanities, social sciences, and natural sciences. Its activities are supported by the UC Press Foundation and by philanthropic contributions from individuals and institutions. For more information, visit www.ucpress.edu.

California Natural History Guide Series No. 91

University of California Press
Berkeley and Los Angeles, California

University of California Press, Ltd.
London, England

Library of Congress Cataloging-in-Publication Data
Russo, Ron.
    Field guide to plant galls of California and other Western states / Ron Russo.
        p. cm. — (California natural history guides ; 91)
    Includes bibliographical references and index.
    ISBN-13 978-0-520-24885-4 (cloth : alk. paper), ISBN-10 0-520-24885-6 (cloth : alk. paper)
    ISBN-13 978-0-520-24886-1 (pbk. : alk. paper), ISBN-10 0-520-24886-4 (pbk. : alk. paper)
    1. Galls (Botany)—California—Identification. 2. Galls (Botany)—West (U.S.)—Identification. 3. Gallflies—California—Identification. 4. Gallflies—West (U.S.)—Identification. 5. Gall midges—California—Identification. 6. Gall midges—West (U.S.)—Identification. I. Title. II. Series.

    SB767.R86 2007
    632'.209794—dc22   2006009332

Manufactured in China
10   09   08   07   06
10   9   8   7   6   5   4   3   2   1

Cover: Speckled gall wasp gall *(Besbicus mirabilis).* Photo by Ron Russo.

The publisher gratefully acknowledges the
generous contributions to this book provided by

the Gordon and Betty Moore Fund
in Environmental Studies
and
the board of directors and members
of the California Oak Foundation
(www.californiaoaks.org)

**California Oak
Foundation**

CONTENTS

ACKNOWLEDGMENTS
AND PREFACE

This major revision of my previous book could not have been completed without the continual assistance of several scientists, colleagues, and friends. I am indebted to John Tucker (oaks) of University of California at Davis; Steven Heydon (cynipid collection) of University of California at Davis; Richard D. Goeden (tephritid flies) of University of California at Riverside; David Headrick (tephritid flies) of California Polytechnic State University; Jim Wangberg (tephritid flies) of University of Wyoming; Patrick Abbot (aphid DNA) of Vanderbilt University; Gwendolyn Waring (gall midges) of University of Arizona; Carl Olson (chalcids) of University of Arizona; Jeffrey Joy (gall midges) of Simon Fraser University; Joseph D. Shorthouse (cynipid wasps) of Laurentian University, Ontario; Robert Zuparko (cynipid collection) of California Academy of Sciences; Bradford Hawkins (gall midges) of University of California at Irvine; Don Miller (manzanita aphids) of California State University at Chico; and E. Durant McArthur (Great Basin shrubs and associated insects) of USDA Forest Service Rocky Mountain Research Station for their generous assistance. I am especially grateful to Raymond Gagné (gall midges) of U.S. Department of Agriculture; D. Charles Dailey (cynipids) of Sierra College at Rocklin; and Kathy Schick (cynipids) of University of California at Berkeley for their critical review of the manuscript and assistance in identifying gall insects; Jerry Powell (moths) of University of California at Berkeley; and Robert Raabe (forest pathology) of University of California at Berkeley for their generous assistance in identifying gall organisms and reviewing appropriate sections of the manuscript. Several people have contributed specimens and encouragement for this work: Steve Edwards and Joe Dahl (East Bay Regional Park District Botanic Garden), May Chen (Audubon Canyon Ranch), Laura Baker, Celia Ronis, Ken Burger, Nan Olson, and Carol Weiske. A special thanks goes to Phyllis Faber who encouraged me to complete this work. Finally, I owe so much to my wife, Sheri, not only for her tremendous help with the manuscript, but also for her hard work as an acute observer, finding many plant galls during our collecting trips.

On a single blue oak in 1970, I found over 30 species of what I later discovered were cynipid wasp galls. Thus began a journey that would take me, over the ensuing years, through every plant community in California and ultimately the western United States. I have found galls nearly everywhere I looked. Immediately apparent throughout all of my searching was the startling fact that such extraordinary creations were so abundant, yet so collectively ignored.

This revision to my previous book, *Plant Galls of the California Region* (1979), represents a much more thorough field guide to galls induced on native and some ornamental plants. This revision not only covers a broader region, but includes over 200 species not in the original book. I have worked to extract as much biological information from the literature—information that has to date been relatively hidden in obscure places inaccessible to most. Every species described in this guide has been collected and recorded by me. While this revision encompasses over 300 species of plant galls, including 95 species on oaks, it still remains incomplete. For so many plant galls, the identities and biology of the causative organisms remain a mystery. Over thirty new species of gall organisms were discovered during the preparation of this guide. Each time I explore a new area, I find undescribed species. You may also find galls not described in this guide or any other reference. Please do not let that frustrate or discourage you. The world is full of discoveries yet to be made. If nothing else, this single idea encourages me, and I hope you, to keep looking and exploring.

Ron Russo

The act of discovery
is one of the greatest joys;
it is a bright flash in the darkness,
with a reward not soon forgotten.

—H. E. EVANS (1968)

IF YOU HAVE EVER CAMPED or hiked beneath oak trees or have oaks on your property or in the neighborhood, you probably have noticed peculiar swellings known as "oak apples" or mysterious colorful adornments on leaves or branches. Or you may have noticed projections, swellings, or pouches on the leaves of alder, willow, juniper, pine, manzanita, sage, creosote bush, wild plum, wood rose, rabbitbrush or any number of native shrubs and trees in the western states. Welcome to the realm of plant galls.

Plant galls are generally produced by host plants in response to the mechanical and/or chemical stimuli of invading organisms. In some cases, these galls are induced by organisms as nursery sites for their offspring, as in the case of mites, psyllids, aphids, thrips, moths, midges, fruit flies, and wasps. In other cases, generally involving fungi and bacteria, invading organisms irritate the host plants through their normal biological functions, resulting in swellings or other disfigurations.

The world of plant galls is a Lilliputian realm in which an organism, sometimes scarcely the size of a period on this page, can induce a plant to produce a swelling of a specific size, shape, and color. Some of these galls, especially those of cynipid wasps, are so flamboyant in design and color they would challenge the wildest of architectural dreams. In their own right, gall organisms are nature's own miniature architects. Part of the "magic" of this little-known world is that plant galls are all around us: in forests, woodlands, marshlands, neighborhood parks, and even your garden.

Along the coastal range mountains of the Pacific states, in the Great Central Valley, in the Sierra and the Cascades, in the arid regions of eastern Oregon and Washington, in the high desert country of the Great Basin, across the Rocky Mountains from Wyoming to Arizona, the diversity of plant species hosts a dazzling array of galls. Many gall-inducing organisms have been described and named by scientists. Many others, however, are new species yet to be discovered, studied, and classified.

Plant galls are known for their incredible variety of shapes, sizes, and colors. They range in size from that of a pinhead on a leaf to a cankerous gall on the side of a tree trunk that exceeds 1.2 m (4 ft) in diameter. In color, they duplicate the full spectrum of the rainbow and then add blends of colors and patterns that dazzle the mind. Some galls are smooth and round. Others have wartlike bumps, spines, hairs, **TUBERCLES**, or flared edges. Some look like balls, saucers, cups, bowls, sea urchins, caterpillars, spindles, clubs, teeth, donuts, or exploded twigs or buds. Some are quite noticeable because of their size or color. Others are nearly impossible to see because they look like normal buds or are so small they escape detection.

Of all the galls and their hosts I have studied over the years, the blue, valley, and Oregon oaks of California and the white oaks of Utah and Arizona produce the most spectacular and flamboyant galls in shape and color. Few insect-induced galls in North America, and perhaps anywhere else, can rival these "architectural wonders."

The purpose of this field guide is to introduce the reader to the causes, effects, and interrelationships of some of the galls and their inducers found in nearly all of the western states. The book is divided into two major sections. The first covers the general biology of galls and **GALL INDUCERS**, host plant effects, and broad ecological relationships with parasites and predators. The second, and largest, portion of this book covers the identification of gall-inducing organisms based on host plants.

Out of necessity, this guide represents a selection of galls found in the region and is by no means a comprehensive representation of existing galls. I am certain that I will continue to find new galls, as will you, that are not described here. While I have collected galls from every one of the western states, the bulk of my work over the last 20 years has been in California, Nevada, Oregon, Washington, and Alaska. Nevertheless, the information in this guide certainly applies throughout the West. Some native plants, for example, have a natural range that extends well beyond California into several other western states. Douglas-fir, madrone, several pines, aspens, cottonwoods, Great Basin sage, rabbitbrush, and creosote brush, among others, have broad natural ranges and likely host the galls described herein throughout their ranges in areas I was unable to visit.

This is not a book on insects, nor is it a book on plants. Instead, this is a guide that attempts to interpret a highly complex and evolved collection of rather challenging interrelationships between plants and plants, between plants and animals, and between animals and animals. In some instances, the galls that result from these highly refined relationships will capture the fancy of anyone who dares to focus attention on even a single, bizarre, brightly colored specimen. In taking time to study just a few such galls, you risk being captivated by the same mysterious fever that has lured countless other enthusiasts to birds, wildflowers, and mushrooms. It only takes a dash of curiosity to explore, discover, and begin to understand the fascination of these strange creations, and, in so doing, the grand biodiversity of Earth itself.

About This Guide

Successful use of this field guide requires that you have a basic knowledge of the taxonomy of native plants (some galls that occur on ornamental plants are not described herein but are only referred to in table 15). In effect, you must know or be able to identify at least the genus, and in some cases, the species, of the host plant to identify the gall organisms described. A number of fine field guides will help you identify plants in the western states. I have systematically used *The Jepson Manual: Higher Plants of California* (Hickman 1993) for taxonomic issues. With some knowledge of plant identities, you can then go to the appropriate group or species and locate the gall(s) of interest. This guide provides as much detailed biological information as is available on each gall species covered, along with host plant and range information. Additionally, several tables list gall organisms known but not necessarily described within the text.

Unnamed Species

I have chosen to include descriptions of the galls of several species not yet classified. Even though I know these galls are induced by gall midges, moths, or cynipid wasps, they are listed as "undescribed." For the most part, these new species are found on oaks (*Quercus* spp.), but several others are found, for example, on huckleberry (*Vaccinium* spp.), gumweed (*Grindelia* spp.), and rabbitbrush (*Chrysothamnus* spp.), in addition to several other hosts. I am placing them in this guide to record their existence, because you are likely to find them, and once taxonomists clarify their identity, you can simply write their names in the proper place.

I have found over 30 species of galls new to science during the preparation of this guide—galls that have not been formally studied and classified. While some of these species are simply listed as "undescribed," in many other cases I have been able to confirm the genus of some of these insects, with the generous assistance of taxonomists. At times identifying species was challenging, and the process of discovering new and unrecorded species reaffirms the notion that there are simply more gall organisms than we know about.

Finally, as part of the preparation of this guide and to confirm questionable species, I visited and studied the cynipid wasp collections at the Bohart Museum of Entomology, University of California at Davis; the Essig Museum of Entomology, University of California at Berkeley; the Museum of Natural History, Sierra College at Rocklin; and the California Academy of Sciences, San Francisco.

What Are Galls?

Plant galls represent just one facet, one platform, a single venue among tens of thousands of ways that plants and animals interact with each other in the world. By the very nature of the existence of gall organisms, thousands of other creatures are able to survive in the complicated, intricate, interdependent existence that drives life on this planet. The world of plant galls is but one among countless stages that exist where the "actors" perform specific roles in an ecological web that supports the rich biodiversity of our planet. As you will discover in this guide, their existence is important if for no other reason than they provide food and shelter to so many creatures.

Galls are tumorlike growths of plant tissue produced by the host plants in response to the chemical and/or mechanical stimuli of invading organisms (fungi, mites, insects), resulting in accelerated production of plant growth hormones (auxins, cytokinins, gibberellins, etc.). Galls are composed of cells that have undergone multiplication in either abnormally high numbers or developed greater size than normal, or plant organs whose growth and development have been altered into forms not otherwise found on the host plants, as is the case of some acorn, bud, and flower galls. The exact mechanism of gall formation, however, may vary widely from one group of gall inducers to another. There are so many variables, requirements, and circumstances, that no one universal method can apply to all gall formers.

Since the galls of many insects (especially wasps and many flies) are specific to their species in size, shape, and color, there is most likely some genetic programming relationship between the compounds provided by the adults or **LARVAE** and the manifestations of plant cellular tissue as influenced by the host's own hormones. Something in the chemicals provided by gall organisms directs the expression of normal plant genes in the development and expansion of the host plant's tissues. Scientists have been looking for a long time for this "blueprint" that seems to control gall characteristics. This has become the "Holy Grail" of gall research.

Based on what is known, at least of cynipid wasps, **MERISTEMATIC** and **PARENCHYMA** tissues are often involved in gall formation. It has been suggested that some gall wasps can convert relatively differentiated tissues back into meristematic tissue. This dedifferentiation of host tissues prevents the normal expression of host characteristics. One aspect of this controlling influence is in the development of the concentric rings of tissue that surround the wasp larval chamber, which is usually lined with parenchymous nutritive tissue on which the larvae later feed.

References: Shorthouse 1973a,b; Weis et al. 1988; Gassmann and Shorthouse 1990; Shorthouse and Gassmann 1994; Brooks and Shorthouse 1997, 1998a,b; Stone et al. 2002; Schick and Dahlsten 2003; Wool 2004.

The Science of Gall Study

While the exact nature of galls has been the subject of much speculation and myth for centuries, the science and biology of plant galls did not begin to develop until the seventeenth century. The science of gall study is divided into two separate fields: the study of plant galls and the insects that induce them, called **CECIDOLOGY**, and the study of how fungi and bacteria gall plants, which falls into plant pathology.

Cecidology brings together entomology, botany, and parasitology in close association. Gall-inducing insects are usually extremely selective in choosing their host plants. Because of this, you can often identify the kind of plant by the species of gall organism, or the gall inducer by the species or group of host plant(s). Furthermore, galls are often so specific in shape, size, and often color to the species of gall inducer (especially with wasps), you can often identify the causative organism without ever seeing it. The scientific names attached to the photos and illustrations of galls in this guide are actually the names of the organisms that induce them.

A Brief History of Galls

Some galls have been well known to industrial and agricultural interests as sources of tanning agents (tannin and gallic acid), printing ink, supplementary livestock feed, and in one case, the cause of major orchard damage (read about crown gall in the section on bacteria that induce galls), but their history goes well beyond modern times. Evidence in the fossil record shows that fungi-induced galls existed 200 to 300 million years ago, during the Upper Paleozoic–Triassic Period in England. Suspected insect-induced galls existed about 225 million years ago, during the Triassic Period in France. The oldest confirmed insect-induced galls from North America are from the Late Cretaceous, about 115 million years ago, and were taken from fossil beds in Maryland. The oldest known cynipid galls are from the Late Eocene, about 34 million years ago, in Florissant, Colorado. Plant galls and gall organisms represent ancient relationships in the evolution of our planet. Today, many of these gall organism–host plant relationships reveal sophisticated and highly evolved organization and development.

While it seems clear that galls and gall organisms are an integral part of the natural landscape, they also played a role in sustaining Native Americans. Evidence exists that Native Americans relied on some galls for specific uses. Several tribes located in California's Central Valley ground the galls of the California Gall Wasp *(Andricus quercuscalifornicus)* into a powder used for making an eyewash and treating cuts, burns, and sores. The "raspberry gall" (in this guide referred to as the sea urchin gall, caused by the Urchin Gall Wasp *[Antron quercusechinus]*) was also ground for use as an eyewash for inflamed eyes. Other galls were ground for use in dyes and hair coloring. The native people from the British Columbia coast reportedly ate the fungus galls found on false azalea *(Menziesia ferruginea)*. It has also been reported that certain spring galls, loaded with tannic acid, were chewed to clean teeth. Native Americans from Arizona reportedly smoked the large round galls most likely caused by the Creosote Stem Gall Midge *(Asphondylia auripila)* found on creosote bush *(Larrea tridentata)*. No doubt there were many other uses for galls that we may never know about.

One of the first people to write about galls was the Greek naturalist Theophrastus (372–286 B.C.), who wrote about the famous gall nuts of Syria. Pliny, the Roman naturalist (A.D. 23–79), recorded 23 medical remedies made with plant galls, including a hair restoration product. For centuries, the production of figs for animal and human consumption has relied on tiny gall wasps. People depended on gall wasps to pollinate the flowers of fig trees (*Ficus* spp.), without understanding the level of intimacy between the different trees and several species of wasps dependent on figs. This is a symbiotic relationship, where the figs can be pollinated only by these gall wasps, and the wasps use only certain species of figs for their floral galls.

In 1861, over 800 tons of "Aleppo" galls (induced by a cynipid wasp, *Cynips gallaetinctoriae*, on a European oak, *Quercus infectoria*) were imported into England for commercial use. Aleppo galls primarily from Turkey were the most-used galls in medicine for many years. In America, over 550,000 pounds of galls were imported from Turkey as late as 1945 for commercial use. Later, a Texas gall, the "mealy oak gall," became commercially popular due to its high tannic acid level of 40 percent. Galls have been used for dyeing wool, leather, and fur and for skin tattooing in East Africa. Galls have even been used as supplemental livestock feed in Missouri and Arkansas *(Dryocosmus deciduus)*. In Mexico, large oak galls *(Disholcaspis weldi)* were occasionally sold in fruit stands because of their reported sweetness.

Because of the lasting quality of the ink produced from European oak galls, gall-based ink has been used for centuries. Monks used gall-based ink in the translation of manuscripts nearly 1,000 years ago. Later, gall-based ink became the preferred ink used by the United States Treasury, the Bank of England, the German Chancellery, and the Danish government, among others. Many important and well-known treaties were signed with gall ink, including the treaty with Japan formally ending World War II in 1945.

Plant galls have a long-standing place in the evolution of our landscape and our human history, and what we know has just scratched the surface of their importance to nature and the well-being of humankind.

References: Swanton 1912; Barrett and Gifford 1933; Balls 1962; Felt 1965; Hutchins 1969; Bean and Saubel 1972; Chestnut 1974; Goodrich et al. 1980; Zigmond 1981; Collier and Thalman 1991; Shorthouse and Rohfritsch 1992; Pojar and MacKinnon 1994; Phillips and Comus 2000; McGavin 2002.

Where Galls Form

Galls can form on every plant part. They occur on roots, trunks, branches, buds, flowers, fruits, and leaves. While some plants such as oaks support galls on all of these parts, other plants such as sage (*Salvia* spp.) and manzanita (*Arctostaphylos* spp.), may support galls only on their leaves. In one analysis in Europe involving oaks, a scientist found that of the cynipid wasp galls known, 5 percent were on roots, 5 percent on buds, 22 percent on branches, 2 percent on flowers, 4 percent on acorns, and about 62 percent on leaves. He also reported that over 80 percent of the galls on members of the rose family (Rosaceae) developed on leaves. The reason for such a disproportionately high gall incidence on leaves is because the leaf is the part of the plant that undergoes the highest metabolic activity during a relatively short growth period. Galls can, therefore, develop relatively quickly in spring and summer. The normal photosynthetic activity of leaves contributes greatly to the rapid development and nutrition of galls and the larvae growing within them.

As you go through this guide, you will notice that some species and groups of host plants support more gall inducers than other plants. Seventy-three percent of the gall inducers in this guide are supported by 16 host plant species or groups (table 1). When you ponder the enormous populations of these species and their associated **INQUILINES, PARASITES,** and **HYPERPARASITES,** the ecological ramifications and importance of such hosts as creosote bush, wild roses (*Rosa* spp.), sagebrush (*Artemisia* spp.), oaks, and willows (*Salix* spp.), among others, are staggering.

References: Mani 1964; Felt 1965.

TABLE 1   Distribution of Gall Species among
Major Host Plants

Host Plant	Number of Species
Alder (*Alnus* spp.)	6
Buckbrush, California lilac (*Ceanothus* spp.)	5
Cottonwood, aspen, poplar (*Populus* spp.)	14
Coyote brush, desert broom (*Baccharis* spp.)	9
Creosote bush (*Larrea tridentata*)	15
Juniper (*Juniperus* spp.)	5
Oak (*Quercus* spp.)	95
Pine (*Pinus* spp.)	7
Plum, choke-cherry (*Prunus* spp.)	5
Rabbitbrush (*Chrysothamnus* spp.)	13
Ragweed (*Ambrosia* spp.)	6
Rose (*Rosa* spp.)	7
Sagebrush (*Artemisia* spp.)	10
Saltbush (*Atriplex* spp.)	8
Service-berry (*Amelanchier* spp.)	7
Willow (*Salix* spp.)	15
Total	227

Note: Includes host plants with five or more gall inducers. Sixteen host plants or
plant groups support 73 percent of the galls described in this field guide.

Gall Inducers at a Glance

For the purposes of this guide, a broad spectrum of galls that develop as a result of the biological activity of invading organisms
such as bacteria, rusts and sac fungi, and mistletoes is covered.
These galls are swellings from which reproductive products or
agents are released. Some regard the swellings that result from
invasion of parasitic mistletoes (*Arceuthobium* spp.) as a reaction
to irritation and not true galls. Mistletoe-induced **WITCHES'
BROOMS,** however, are included as galls in this guide. Also included is a dazzling array of galls that are induced by insects or
mites to feed and produce offspring, or by their larvae, to house
and nourish themselves as they grow.

The Evolutionary Path

During the millions of years of evolution and adaptation among insects, natural selection influenced the biological paths that each group and species followed. As a result, each group and species ultimately found specific niches within which they operated for the purposes of survival and procreation. Some insects became wood borers, pollinators, or carrion-eaters, while others rolled dung, captured spiders, or parasitized other animals. Somewhere along this evolutionary line many species of insects and mites evolved intimate relationships with plants, resulting in the development of galls. Like other insects and mites, the evolving gall inducers became extreme specialists. Worldwide, there are about 13,000 species of gall-inducing arthropods. Among them there are at least 21 separate groups, but for our purposes we shall consider only eight: eriophyid mites (Eriophyidae), psyllids (Psyllidae), aphids (Aphididae) and adelgids (Adelgidae), beetles (Cerambycidae), moths (Gelechiidae, Tortricidae, Cosmopterigidae), flies (leaf-mining flies [Agromyzidae], tephritid fruit flies [Tephritidae], and gall midges [Cecidomyiidae]), and wasps (chalcids [Chalcidae], sawflies [Tenthredinidae], and gall wasps [Cynipidae]). The United States has over 2,000 known species of gall-inducing insects. Nearly 1,000 of these species are cynipid wasps, with an additional small group of sawflies. There are also about 800 species of gall midges. Next in prominence are the tephritid fruit flies. While the exact number of gall-inducing tephritids in North America is not clear, several species are common in the West. Eriophyid mites are responsible for a considerable number of galls, while fewer aphids and moths induce galls. There are, however, far more moths that induce galls in the West than are listed in this guide. At least two leaf-mining flies and two chalcid wasps are known to gall western plants. Only one gall-inducing beetle is listed here, even though many beetles are inquilines in galls. The only psyllids included here occur on ornamentals (see table 15).

The second group of gall inducers includes numerous species of bacteria, fungi, and mistletoes that stimulate the production of root, stem, and leaf galls. Unlike some insect galls, these agents often produce much less specificity in the size, shape, and color of galls, making identification using these characteristics difficult and sometimes impossible. For some, identification is possible based on the identity of the host and the location and type of gall. Some bacterial galls (nitrogen-fixing nodules) appear on the roots of host plants such as lupines (*Lupinus* spp.), alders (*Alnus* spp.), and the group of shrubs generally referred to as California lilac (*Ceanothus* spp.). These galls are rare examples of situations where the host plant and the gall inducers mutually benefit. Other bacterial galls appear just below the soil surface of fruit trees and ornamental roses. Fungus galls are associated with stems, leaves, flowers, and fruit (cottonwood [*Populus* spp.], alder, choke-cherry and plum [*Prunus* spp.], and wood fern [*Dryopteris arguta*]). Mistletoes almost always induce stem swellings and sometimes witches' brooms on host plants (pines [*Pinus* spp.], firs [*Abies* spp.], incense-cedar [*Calocedrus decurrens*], among

	NUMBER OF SPECIES	
	On Native	On Ornamental
AGENT	Plants	Plants
Bacteria	3	2
Fungi		
Rust	5	
Sac	10	1
Mildew	1	
Mistletoes (Viscaceae)	5	
Mites (Eriophyidae)	25	3
Aphids (Aphididae and Adelgidae)	13	
Psyllids (Psyllidae)		2
Moths (Gelechiidae, Tortricidae, Cosmopterigidae)	11	
Beetles (Cerambycidae)	1	
Leaf-mining flies (Agromyzidae)	2	
Tephritid fruit flies (Tephritidae)	12	
Gall midges (Cecidomyiidae)	106	1
Chalcid wasps (Chalcidae)	2	
Sawflies (Tenthredinidae)	8	
Cynipid wasps (Cynipidae)	98	2
Total	302	11

TABLE 2 Gall-Inducing Species Covered in This Guide

Note: Does not include species referred to within a species' description.

others). See table 2 for a listing of the gall-inducing organisms covered in this guide.

Another group of gall-like structures appears on trees such as buckeyes (*Aesculus* spp.), California bay *(Umbellularia californica)*, coast redwood *(Sequoia sempervirens)*, and the ornamental pepper tree *(Shinus molle)* and are actually genetic anomalies or simply adventitious buds, not true galls resulting from invading organisms. See the "Redwood False Galls." The biology of each major group of gall inducers is discussed in a later section.

References: Mani 1964; Felt 1965; Sinclair et al. 1987; Johnson and Lyon 1991; Shorthouse and Rohfritsch 1992.

Common Types of Galls

Most galls occur on leaves, **PETIOLES**, stems, and branches. Leaf galls are divided into specific types based on their general structure. Leaves support roll galls, fold galls, filzgalls **(ERINEUM)**, pouch and bead galls, and mark or spangle galls. On stems and branches, we see either **INTEGRAL** swellings or detachable outgrowths. Two other forms of galls include **FASCIATIONS** and witches' brooms. Other swellings, usually at the base of a tree trunk or high along the main trunk, are called "burls" and are often mistaken for galls.

ROLL GALLS: These galls are characterized by the outer edge of the leaf rolling inward, encompassing the gall organism (fig. 1). The rolled leaf tissues are either slightly or noticeably swollen and may be much harder than surrounding tissues. Mites, moths, and flies generally produce roll galls.

FOLD GALLS: These galls involve either the outer edge of the leaf simply folding over one time inwardly to encompass the gall organism, or the leaf folding along the **MIDRIB** vein, creating a pouch on one side of the leaf (fig. 2). Swelling and distortion of leaf tissues are also characteristic of these galls. Aphids, moths, and flies usually induce fold galls.

FILZGALLS: Filzgalls (erineum pockets or galls) are hair-lined pockets or depressions (fig. 3). These pockets are usually noticeable on one side of the leaf. Eriophyid mites usually induce erineum galls. A fungus (*Taphrina* sp.) is also known to create a filzgall-like structure on cottonwood leaves.

POUCH AND BEAD GALLS: These galls are either small, round or globular beads, or resemble clubs or spikes on the leaf surface (fig. 4). Both forms have an opening at the base, which allows escape of the mites or flies when mature. Eriophyid mites and some gall midges induce pouch and bead galls.

MARK AND SPANGLE GALLS: In most cases these leaf galls completely enclose the insect with no openings, the exception is some cecidomyiid galls. Mark and spangle galls are either detachable or are an integral part of the leaf, and both tend to be the most flamboyant in color and shape (fig. 5). Cynipid wasps and a few cecidomyiids induce most mark and spangle galls on leaves (some cecidomyiid galls have openings leading to the larval chambers, but others do not).

FASCIATIONS: In these galls, the terminal buds of numerous plants are stimulated to fan out, creating extraordinarily flattened and rather striking shapes, often resembling elk or moose antlers (fig. 6). Sometimes, mites or fungi induce these structures. The majority of fasciations tend to be localized genetic abnormalities and are usually referred to as "noninfectious fasciations." Fasciations occur commonly on ornamental plants, as well as a few native species (table 3).

WITCHES' BROOMS: These are common on conifers but also occur on many native shrubs (table 4). Brooms usually involve a dense collection of small branches and shoots emanating from a common focal point (fig. 7). Often these die after one season.

TABLE 3   Partial List of Host Plants with Fasciations

Aloe (*Aloe* spp.)

Australian brush-cherry (*Eugenia myrtifolia*)

Chocolate lily (*Fritillaria lanceolata*)

Cotoneaster spp.

Cryptomeria spp.

Evening primrose (*Oenothera* spp.)

Mugwort (*Artemisia douglasiana*)

Mullein (*Verbascum thapsus*)

Poison oak (*Toxicodendron diversilobum*)

Sagebrush (*Artemisia* spp.)

Saguaro (*Carnegiea gigantea*)

Note: The saguaro, from Arizona, is a single plant that has become famous, attracting photographers from throughout the West.

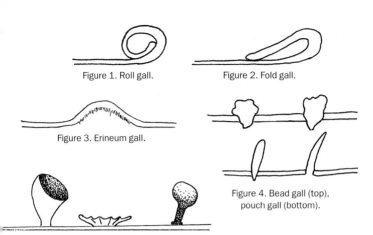

Figure 1. Roll gall.

Figure 2. Fold gall.

Figure 3. Erineum gall.

Figure 4. Bead gall (top),
pouch gall (bottom).

Figure 5. Mark or spangle galls (detachable).

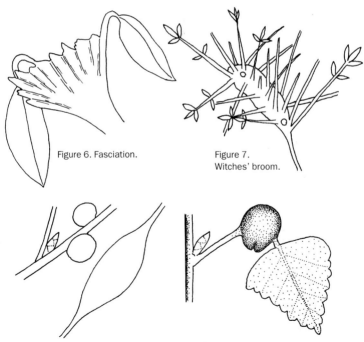

Figure 6. Fasciation.

Figure 7.
Witches' broom.

Figure 8. Detachable stem gall (left), integral stem gall (center),
integral petiole gall (right).

TABLE 4   Witches' Broom Galls

Agent	Host Plant
Mildew	
Live Oak Witches' Broom Fungus (*Cystotheca lanestris*)	Coast live oak (*Quercus agrifolia*)
Mycoplasmalike	
Undescribed	Willow (*Salix* spp.)
Rust Fungus	
Baccharis rust gall fungus (*Puccinia evadens*)	Coyote brush (*Baccharis pilularis*)
Chrysomyxa arctostaphyli	Colorado blue spruce (Arizona, Utah) (*Picea pungens*)
Gymnosporangium confusum	Juniper (*Juniperus* spp.)
G. nidus-avis	Juniper
Huckleberry broom rust (*Pucciniastrum goeppertianum*)	Huckleberry (*Vaccinium* spp.)
Incense-cedar rust (*G. libocedri*) *Melampsorella caryophyllacearum*	Incense-cedar (*Calocedrus decurrens*) Fir (*Abies* spp.) (several species)
Western gall rust fungus (*Endocronartium harknessii*)	Pine (several species)
Sac Fungus	
Apiosporina collinsii	Service-berry (*Amelanchier* spp.)
Elytroderma deformans	Pine (several species)
Leaf gall fungus (*Exobasidium vaccinii*)	Manzanita (*Arctostaphylos* spp.), Rhododendron (*Rhododendron* spp.)
Service-berry fungus (*T. amelanchieri*)	Service-berry (*Amelanchier* spp.)
Taphrina aesculi	Buckeye (*Aesculus* spp.)
T. thomasii	Holly-leafed cherry (*P. ilicifolia*)
Witches' broom fungus (*T. confusa*)	Choke-cherry (*Prunus virginiana*)
Virus	
Nanus holodisci	Ocean spray (*Holodiscus discolor*)
Mistletoe	
Arceuthobium americanum	Lodgepole pine (*Pinus contorta* subsp. *murrayana*)
A. apachecum	White pine (Arizona) (*P. strobus*)
A. californicum	Sugar pine (*P. lambertiana*)
A. campylopodum	Coulter pine (*P. coulteri*), Jeffrey pine (*P. jeffreyi*), knobcone (*P. attenuata*), ponderosa pine (*P. ponderosa*)

A. cyanocarpum	Limber pine *(P. flexilis)*, western white pine *(P. monticola)*
A. laricis	Larch (*Larix* spp.)
A. occidentale	Bishop pine *(P. muricata)*, Coulter pine, Monterey pine *(P. radiata)*
A. tsugense	Hemlock (*Tsuga* spp.)
A. vaginatum	Ponderosa pine
Douglas-fir dwarf mistletoe *(A. douglasi)*	Douglas-fir *(Pseudotsuga menziesii)*

Note: Includes many of the witches' brooms known to occur in the western states.

Large brooms exceeding four feet in diameter have been found on Douglas-fir *(Pseudotsuga menziesii)*. Witches' brooms usually involve bacteria, fungi, mistletoes, or mites. Broomlike clusters of shoots that originate from external mechanical injury not associated with an internal organism are not true galls. For example, porcupines have the habit of repeatedly chewing the same area of a conifer trunk, which results in a broomlike cluster of shoots. From a distance, these dense clusters of branches might look like any other normal mistletoe-induced witches' broom.

STEM AND BRANCH GALLS: These galls involve a variety of causative agents including rust fungi, mistletoes, flies, moths, beetles, and wasps. The galls are either integral and nondetachable or protrude from the branch and are detachable (fig. 8). Integral stem galls can disrupt the flow of nutrients to outer regions, thereby killing the branch beyond the gall. Common curios from South America sold in tourists' shops in California and elsewhere are insect galls carved into lizards and birds (pl. 1). These galls look like

Plate 1. Curio imported from Central America, probably caused by a chemical interaction between the host branch and a gall wasp.

an exploded branch, creating a fan-shaped natural art form, which allows easy escape of the adult wasps.

BURLS: These are knobby outgrowths with unusual grain, often at the bases of trees. Redwood burls are often sold as curios or for wood furniture. Even though burls have been referred to as nothing more than trunk galls, they usually do not involve invading organisms that stimulate their development. Rather, burls are the manifestations of dormant buds or adventitious buds and, therefore, are not true galls. California bays, redwoods, buckeyes, and pepper trees often sport knobby swellings or burls along their bases.

CANKERS: Several viruses, bacteria, and fungi induce massive eruptions and swellings of trunks and branches. These eruptions do not always have a gall-like form, although some are fissured swellings that ooze sap. For the purpose of this guide, most cankers on trees have been ignored except for those prominent swellings induced by certain bacteria and rust fungi.

ROOT NODULES: A number of native plants have evolved a symbiotic relationship with nitrogen-fixing bacteria. Many of these plants tend to grow in nutrient-poor soils, particularly where the presence of nitrogen is low. As a result of this relationship, nodules form on the taproot, root hairs, and lateral roots that concentrate nitrogen from the air in usable form. In effect, these nodules are bacterial root galls and are considered as such in this guide. The nodules are created by nitrogen-fixing bacteria of the genera *Frankia* or *Rhizobium*. Ultimately this nitrogen becomes available to other plants, thus allowing establishment of plants in

TABLE 5   Nitrogen-fixing Bacteria and Host Plants

Bacteria	Host Plant
Frankia spp.	Alder (*Alnus* spp.)
	Antelope brush *(Purshia tridentata)*
	Bayberry *(Myrica pensylvanica)*
	Buffalo berry (*Shepherdia* argentea)
	Ceanothus (*Ceanothus* spp.)
	Mountain-mahogany (*Cercocarpus* spp.)
	Wax myrtle *(Myrica californica)*
Rhizobium spp.	Most legumes (includes lupine [*Lupinus* spp.])

terrain otherwise unsuitable. See table 5 for the species on which these nodules are known to occur.

References: Mani 1964; Felt 1965; Meyer 1987; Sinclair et al. 1987; Johnson and Lyon 1991; Walls and Zamora 2001.

Seasonal Appearance and Growth Rate

The appearance of plant galls in nature is influenced by many factors but generally coincides with the season of greatest plant growth. For the most part, galls develop on their host plants between spring and late summer. Some midge and fruit fly galls, however, develop on rabbitbrush during winter.

Generally, the insects that are successful are those whose emergence (from old galls or the duff under their host plant) coincides within a week or so of the development and optimal condition of their preferred host's gall organ (swelling buds, new shoots, and leaves). If they emerge too early or too late, they may miss the optimal period for reproduction and egg laying. Some insects remain in **DIAPAUSE** (prepupal sleep) for a year or more before they complete metamorphosis into adults. While the advantage of producing longer generations is not well understood, multiyear life cycles may represent a mechanism that helps stabilize the population, given the availability of preferred sites and the variability of weather, predation, and the potential for environmental hardships.

In spring and summer, depending on altitude, vast amounts of food energy are directed into the production or expansion of buds, new shoots, branches, and leaves. During these seasons, many galls seem to appear overnight. Actually, their growth requires much more time. In summer at low elevations, most plant growth slows down, yet during this period, when nutrients would normally be used to sustain already existing plant organs, energy can be redirected into the production of a second flush of galls (as seen on oaks with cynipid wasps). The growth of these summer galls normally proceeds at a slower rate than that of spring galls.

An individual blue oak *(Quercus douglasii)* may produce both a spring and a summer crop of galls. Many species of cynipid wasps exhibit two alternating generations that induce different galls often on different plant parts at separate times of the year. This **ALTERNATION OF GENERATIONS** is a characteristic trait of cynipid wasps on oaks and is known as **HETEROGENY**. Usually, these insects have two modes of reproduction. The spring-generation insects, for example, comprises males and females who reproduce sexually, while the summer-fall generation is typically female only. These females reproduce without males (a process called **PARTHENOGENESIS**), resulting in next year's spring galls bearing males and females. For many other gall organisms, there are no known males, leaving the females of these species to reproduce strictly parthenogenetically. This interesting alternation of reproductive modes confused early entomologists, who thought the different galls belonged to different species, not realizing they were alternating generations of the same species.

There are so many diverse species of gall organisms with complicated life requirements, we cannot apply a simple rule of thumb to all. The cynipid and tenthredinid wasps discussed herein are among the most highly evolved and complicated to understand.

References: Weld 1957; Lyon 1959, 1963, 1964, 1969, 1970, 1993; Doutt 1960; Rosenthal and Koehler 1971a; Dailey and Sprenger 1973a,b; Russo 1975, 1979, 1983, 1990; Wool 2004.

Environmental Factors

If you walk through a blue oak woodland or a manzanita chaparral community, you will find some trees and shrubs covered with plant galls, while others nearby appear to lack galls. The reasons for these sharp variations involve many environmental factors, as well as specific microenvironmental requirements of the gall organisms, of which we have little understanding.

Environmental factors such as exposure to the sun and high temperatures, shade, humidity, soil characteristics, and host chemistry may play a significant role in the physiology and stress levels of the host plant and therefore its suitability as an egg-laying site. A willow growing in a warm, southwesterly exposure may not support the diversity and abundance of gall organisms as the same species of willow growing in a cooler, shaded, northeasterly location, or vice versa. The thickness of the leaf epidermis, the presence of high volumes of terpenes, and other chemical and physiological elements may impact the suitability of host plants. A gall insect must not only locate the right species of host plant during its short lifetime, it must find one in the proper physiological condition that is suitable for egg deposition.

Because so many gall insects are tiny and not strong flyers, their success in finding the proper host may be seriously influenced by the condition of the wind on the days these insects emerge. Many questions revolve around wind dispersal and **EDGE EFFECT.** Whether plants at the edge of established stands are more likely to support higher populations of gall insects than those protected from wind in the interior of the community needs further study. However, some assumptions are worth considering. Trees and shrubs upslope and separated from related patches downwind are likely to catch weak flyers blown off course on the day of emergence. Conversely, those interior trees that do support gall inducers are more likely to maintain several generations of the same species who come up from the duff below. Assuming that the conditions for gall development and gall insect survival are suitable, a single tree or shrub can support generation after generation of the same species, regardless of its location. I have found a single, small (less than 4 m [13 ft] tall) blue oak growing alone in the midst of an expansive grassland that supported nine species of cynipids and their galls. These wasp generations most likely used the same host tree year after year, based on the likelihood of their moving directly from the ground up into the original host tree upon emergence, rather than migrating laterally from a neighboring tree some distance away.

Populations of gall-inducing insects may vary dramatically from season to season and year to year depending on a variety of factors. Unusually harsh winters with extreme temperatures, high winds on emergence days, large populations of competitors, parasites and predators, wildfire, high summer temperatures, and drought impacts on the host plant, among other factors that are unknown, may influence population levels within a species.

One example of interspecies competition is seen in the relationship and concurrent cycles of the moth known as the California Oak Worm *(Phryganidia californica)* and the Two-horned Gall Wasp *(Dryocosmus dubiosus)*. Both insects require the leaves of coast live oak *(Quercus agrifolia)*. When the caterpillars of the Oak Moth defoliate live oaks, they eliminate the egg-laying sites—and therefore, galls—for the wasps. About every seven years, when the moth reaches a peak in its population cycle, entire woodland areas can be defoliated. Without leaves, the wasps face serious problems. Conversely, at the peak of the population of Two-horned Gall Wasps, each live oak leaf can bear dozens of galls attached to the midrib and lateral veins on each leaf's underside. Such a presence deprives outer leaf tissues of the nutrients required to sustain photosynthesis, resulting in massive browning of leaf margins, called "scorch." Under these conditions, the low palatability of what is left may impact the feeding success and survival of the moth caterpillars. So, potentially, each species may negatively impact the other. While there are no empirical data on this relationship, I suspect that the peaks in population levels for each species occur when the other is at its lowest level, and there appear to be no data that shows that either population stresses host oaks beyond recovery.

Research on creosote bush in desert regions has shown that plants growing under stressful conditions (less water and nutrients, along with physiological impacts) on talus slopes supported higher numbers of galls and gall-inducing species than those plants growing on desert flats. In this same study, only two of the eight species of gall midges studied were more abundant on nonstressed plants. Stressed plants apparently produced more terminal branches compared to those on flats, where plant growth occurred primarily through the lengthening of the existing branches. This research points not only to the increased availability of potential galling sites as a key factor, but also to the potential of chemistry as a key determinant in gall and gall-inducing species numbers.

References: Harville 1955; Brown and Eads 1965; Russo 1975, 1979; Waring and Price 1990; Williams et al. 2002; Wood et al. 2003.

Damage to Host Plants

Generally, the overall vigor and health of a shrub or tree is not significantly affected by the seasonal production of galls. Localized damage may be sustained if a branch, group of leaves, or a single leaf produces either a large number of galls, such as those of the Two-horned Gall Wasp on coast live oak, or galls of a particular type, such as the witches' brooms of incense-cedar rust *(Gymnosporangium libocedri)* on incense-cedar. Significant damage to a host plant as a result of gall production would depend on several factors. Variables include the size of the host plant, the soil and exposure conditions under which it is growing, and the degree of infestation or infection by a particular gall organism. Severe damage can occur to a weakened tree if it is induced to produce, for example, large numbers of giant oak apples induced by the California Gall Wasp or integral stem galls consistently over successive years. Stunted valley oaks *(Quercus lobata)* can be found growing in nutrient-poor soil and heavily laden with oak apples. Looks can be deceiving, however, because these oak apples may remain on the host tree for three to four years, even though the insects generally complete all growth and activity within the first year. A massive number of these galls on a stunted tree could be the accumulation of several years, making it look as if the tree is severely stressed by the galls, when soil and water or other environmental conditions may be more important influences.

Another cynipid, the Two-horned Gall Wasp, is responsible for damage that draws the attention of homeowners. The galls of this wasp develop on the midrib and lateral veins on the undersides of coast live oak leaves. With a dozen or more of these galls rerouting nutrients within a single leaf into the gall tissues, the outer leaf tissues beyond the galls are deprived and ultimately die, turning brown. As mentioned earlier, in years when the population of this gall wasp is peaking, the coast live oaks look like they are dying. This condition, called scorch, is seasonal, as the live oaks bounce back the following season with new spring leaves. Also see the earlier discussion regarding California Oak Worms in the Environmental Factors section.

A third situation that draws public attention involves a bacterial attack on the young acorns of coast live oak and interior live oak *(Q. wislizenii)* trees that have been galled by either the Live Oak Petiole Gall Wasp *(Callirhytis flora* [syn. *C. milleri]*) or *Eumayria eldoradensis,* both cynipid wasps. The adult wasps puncture the acorns with their OVIPOSITORS to lay eggs. The bacterium *Erwinia quercina* enters the acorns through these holes. After awhile, a frothy ooze develops and begins dripping, landing on everything under the trees including cars and driveways, to the consternation of many people. Without the presence of the bacteria, people would scarcely notice the activities of these acorn gall wasps. What effect the bacteria have on the gall larvae is unknown.

Another example occurs with the common landscape shrub

Australian brush-cherry *(Eugenia myrtifolia)*. This shrub is galled by a psyllid that causes severe leaf deformities and branch-tip fasciations. Galled leaves are usually covered with dark red, bump-like pit galls. Heavy pruning seems to be effective in controlling this gall pest. Even though a plant appears heavily infested, it can live for many years with its psyllid pests.

Another major concern is peach **LEAF CURL** caused by the sac fungus *Taphrina deformans,* which is widespread and results in significant swelling and distortion of leaves and loss of flowers and fruit. The fungus is common among ornamental peaches and nectarines. The leaves are often light green or reddish in color but eventually become completely white due to spore production by the fungus. Another species, the plum leaf-curl fungus *(T. flectans),* induces leaf curl in wild cherries (*Prunus* spp.). Because of the extensive cell multiplication induced by these sac fungi, they should be considered gall-inducing organisms in the West, as they are in England.

Finally, orchardists and nursery workers are quite familiar with **CROWN GALL** associated with the crown gall bacterium *Agrobacterium tumefaciens.* This widespread soil bacterium attacks fruit, nut, and shade trees, as well as commercial roses and other ornamental plants, creating large, swollen galls at the soil surface at the base of trees or shrubs and sometimes on aerial parts of the plants and the root systems. The crown gall bacterium carries a plasmid, which has a gene responsible for inducing the gall. The bacterium itself is not the gall-inducing agent. These galls can exceed a foot in diameter, impair the flow of nutrients, and serve as an opening for damaging fungi.

The Balsam Woolly Adelgid *(Adelges picea)* is responsible for stunting and killing thousands of trees with its gouty terminal bud galls. It is a serious pest on tree farms in the Northwest. Another serious problem involves the Douglas-fir dwarf mistletoe *Arceuthobium douglasii,* which induces massive witches' brooms on Douglas-fir throughout the West. Heavily broomed trees can be weakened and killed.

While these galls and their related secondary invaders can be serious health threats to their host plants, the vast majority of gall organisms in nature cause little long-term damage to their host plants. When you consider that most gall-inducing wasps, midges, and mites use the same host shrubs and trees year after year, taking only relatively insignificant amounts of plant nutrients, and combine that with natural parasite and predator controls, there is little chance of severe damage to host plants.

References: Hildebrand and Schroth 1967; Russo 1979; Sinclair et al. 1987; Johnson and Lyon 1991.

Galls as Nutrient Sinks

There is growing evidence that as galls develop for some species, particularly cynipid wasps, they redirect macronutrients and micronutrients from roots and leaves into gall tissues, creating "nutrient sinks" within the host plant. These sinks contain higher concentrations of several nutrients than are normally found in other nongalled plant tissues of roots, stems, leaves, and fruit. Within cynipid gall tissues, parenchyma cells form a nutritive layer around the larval chamber and sequester both macronutrients (e.g., nitrogen, phosphorus, magnesium) and micronutrients (e.g., iron, manganese, zinc). In fact, several researchers have documented the relocation of carbohydrates, lipids, proteins, and other nutrients into gall tissues. Galled leaves undergo increased photosynthetic rates, with the by-products being intercepted by the galls. In this light, galls are viewed as the delivery mechanisms of superior food resources to larvae (Shorthouse and Rohfritsch 1992), in contrast to the nutritional quality of ungalled tissues that are consumed, for example, by nongalling beetles and other wood borers.

We now know that galls can act as an interceptor or block to the normal flow of resources, but galls might also redirect the translocation of nutrients from various plant parts. In this manner, galls would draw nutrients from tissues beyond those that would normally carry the flow through the attacked organ. The position of the gall on the plant may determine whether the gall acts as an interceptor or redirector. Active concentration of nutrients within galls seems to be higher during the growth phase of galls rather than after the galls have matured and the larvae are further developed.

In Canada, researchers focused on lowbush blueberry (*Vaccinium angustifolium*) growing near an ore smelter. They found that the roots nearest the smelter contained the highest concentrations of copper and nickel, and that the concentrations logarithmically declined with distance from the smelter. When the team looked at blueberries near the smelter galled by the chalcid wasp *Hemadas nubilipennis,* they found that the greatest concentrations of copper and nickel were in the gall tissues and not the roots.

In another Canadian study, researchers found that some gall inducers can apparently regulate the amount of nutrients within gall tissues or at least take only what they need. The galls of the cynipid wasp *Diplolepis triforma* on *Rosa* spp. had greater concentrations of nitrogen and most mineral nutrients than those of *D. spinosa*. Similarly, the larvae of *D. triforma* and the parasites feeding on the larvae contained higher concentrations of nitrogen and other nutrients than those of *D. spinosa*. Interestingly, the parasites of both species were similar in mineral composition to their hosts. Yet, in this case the galls of both species contained less nitrogen and minerals than ungalled tissues on the same hosts. Although this may appear to be in contrast to the nutrient-sink concept, it does suggest that these two species, unlike others in the genus, are mobilizers of their host's resources utilizing only those elements essential to their development.

In yet another study, two researchers looked at the concentration of water in gall tissues from fourwing saltbush *(Atriplex canescens),* a small shrub growing in arid regions of the South-west. They found that certain cecidomyiid galls (*Asphondylia* spp.) concentrated water in the gall tissues at the expense of the stems and branches. Furthermore, they found that shrubs growing near a spring supported significantly higher numbers of galls than those shrubs in a drier habitat, which suggested that the quantity of water available to the gall and the larvae was a prime determinant in the ability of the host plant to support gall midges. Moreover, these results were reversed in a study of roses and the Pincushion Gall Wasp *(D. rosae),* where plants suffering from a lack of water supported more galls.

Clearly, each species may require different levels of water and various minerals and other nutrients to complete their development. Therefore the concentrations of nutrients in galled compared to ungalled tissues may vary widely from species to species and host plant to host plant. The story of these highly complex chemical relationships is only now beginning to unfold, and many questions and years of research remain.

References: Hawkins and Unruh 1988; Bagatto and Short-house 1991; Bagatto et al. 1991; Shorthouse and Rohfritsch 1992; St. John and Shorthouse 2000; Stone et al. 2002; Williams et al. 2002; Wood et al. 2003.

The Gall Community

No plant or animal in nature stands alone. Each is connected to other living creatures in one form or another. The world of even a single gall-inducing insect can be so convoluted and interwoven with its environment it could drive you crazy trying to unravel it. If the galls lying under a blue oak, for example, are not destroyed in winter by fungi and bacteria and not eaten by birds or mice, and the adult insects emerge, they face an onslaught of predatory lizards, spiders, flies, wasps, and birds before they find a suitable site to lay eggs. Some gall insects, especially the wasps, do not eat as adults and live only for a week or so. This increases the pressure on these insects to find suitable egg-laying sites as soon as possible. If a gall inducer makes it this far, having laid eggs, and if the eggs hatch, the larvae and resulting galls can attract: **INQUILINES**, those insects (mostly vegetarian) that eat gall tissues; **PARASITES** that attack the gall-inducer's larvae; and **HYPERPARASITES** that attack the larvae of the parasites. As if all this were not difficult enough, sometimes even the seemingly innocuous inquilines can switch tactics and kill any other insect they encounter within the gall while feeding on gall tissue. And finally, some insects simply prey on whatever insects they find inside the galls.

Over a two-year period, one study showed that the inquiline *Periclistus pirata* accounted for between 55 and 65 percent mortality in the Rose Stem Gall Wasp *(Diplolepis nodulosa)*, a cynipid, on roses in Ontario, Canada. In the same study, six different species of parasites caused only 17 percent of the mortality in the gall inducers, with an additional 13 percent mortality by other inquilines. Consequently, one inquiline species had a much more dramatic impact on the gall inducers than six other parasite species and other inquilines combined. When you add the cumulative effect on the gall-inducer population, the combined loss can reach 80 to 90 percent of the population. These numbers and the suggested impacts vary greatly from one species to another, as well as from one circumstance or environment to another, but their role and importance are staggering.

In another study, researchers found that a wasp, *Tetrastichus cecidobroter,* an inquiline, attacked the galls of the Tumor Stem Gall Midge *(Asphondylia atriplicis)* on the shrub fourwing saltbush (see Saltbush Galls). In so doing, these inquilines create their own galls, called **ENDOGALLS**, within the primary host gall. The developing galls of the inquiline bulge into the larval cavities of the host midges, crushing them. In this rather unusual case, most of the gall midge larval loss is actually due to the presence of the inquiline. "Galls within galls" seems strange, but this is not the only case, as we shall see later with the midge *Rhopalomyia bigeloviae,* which creates galls inside the galls of the Bubble Gall Tephritid *(Aciurina trixa)* on rabbitbrush (see Rabbitbrush Galls).

These strange interactions and relationships paint a picture of the intricacies of survival. Theoretically, even in years of high loss to parasites and inquilines and other environmental factors, enough of the gall inducers survive and mature to continue their species. The inhabitants of galls, whether they are gall inducers, inquilines, parasites, or hyperparasites, can serve as focal points, which, in turn, attract many other predatory insects. The actual number of insects supported by the galls, including the gall inducers and associated insects, can easily exceed several dozen species. Even if we look only at galls, the ecological web of a chaparral or woodland community is a jumble of interactions, dependencies, and implications that far exceeds our wildest imagination.

References: Gordh and Hawkins 1982; Hawkins and Goeden 1982; Brooks and Shorthouse 1997; Headrick and Goeden 1997.

Parasite-Inquiline Influence on Gall Shape

Parasites and inquilines can severely impact the normal development of gall tissues. The timing of insertion and attack, general behavior, and, perhaps, added chemistry of parasites and inquilines can either abort gall development or cause galls to develop in a manner that makes them look like they belong to new species of gall inducers. As you find and collect galls, you will certainly find specimens that do not match the photographs or illustrations in this guide and others. While it may be tempting to assign new species status to such finds, we need to be careful in studying and understanding the exact nature of the complex relationships between gall makers and their associated parasites and inquilines first. As mentioned earlier, all kinds of insects emerge from galls, not just the gall inducers, and their exit holes may look similar. Some examples of inquiline-influenced galls are provided later in this guide (see the gall account for the Beaked Spindle Gall Wasp *[Heteroecus pacificus]*).

Gall-Inducer Defense

Seemingly vulnerable larvae may be protected in a number of ways. Some are discussed here, while yet another mode is discussed in the next section, on **HONEYDEW** and yellow jackets.

Normal plant defenses against herbivorous insects (terpenes, tannins) may be influenced by gall inducers and used to their own advantage. While tannins, for example, are absent from the nutritive tissues near the larval chambers in cynipid oak galls, their levels are usually very high in other gall tissues surrounding the larval chambers. High concentrations of tannins in tissues surrounding the larval chambers may discourage deeper penetration to the chambers by inquilines, thereby protecting the larvae.

Other chemical defenses are employed by gall inducers. Studies have shown that the larvae of the sawfly *Neodiprion sertifer* have the ability to sequester terpenes from their hosts in pouches of the foregut. When disturbed, the larvae regurgitate the terpene compounds as a borrowed defense against attack. A similar behavior was observed with an Australian sawfly, *Perga affinis,* using eucalyptus oils as an orally discharged defense. Sawflies are known for their ability to eject a pungent, vinegary fluid from their **VENTRAL** eversible glands. Sawflies raise and wave their abdomens when disturbed to better expose these "stink glands." Even with such pungent defenses, sawflies often die when their galls are destroyed by gall-tissue-eating inquilines.

A number of mitigating factors influence the degree of success parasites have in locating gall insects and parasitizing them. The thickness of some galls, as in the case of the California Gall Wasp, may prevent a parasitic wasp such as an ichneumon from reaching the larvae with its ovipositor. Also, in **POLYTHALAMOUS** galls (with numerous larvae in separate larval chambers) (fig. 9) the innermost larvae are less vulnerable to parasitic attack than those closer to the gall's surface. In small, **MONOTHALAMOUS** galls (containing one or more larvae in a common chamber) (fig. 10), the larvae are more vulnerable to attack. One of the character-

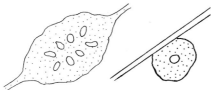

Figure 9 (left).
Polythalamous gall.

Figure 10 (right).
Monothalamous gall.

istics of cecidomyiid galls is that often several larvae occupy a common chamber, which makes them particularly vulnerable to attack because of the ease with which parasites and predators locate victims.

In a study of chalcid parasitoid attacks on cynipid wasp galls *(Acraspis hirta)*, researchers found that the chalcids concentrated their attacks on the host galls while the galls were still fresh, tender, and growing. The softer tissues of developing galls were more conducive to parasite attack than galls that had matured and hardened. Parasite **OVIPOSITION** was significantly inhibited as a result of the hardening of gall walls. Gaining entry under the right conditions for the parasites does not always mean the immediate death of the gall insect's larvae, however. In some cases, even though entry and location of the gall insect larvae occurred during the early stages of gall development, the gall larvae were allowed to develop normally before actual parasite feeding began.

Furthermore, it was found that there were significant differences in parasite success between galls that were spaced out on the host plant to a density of one gall per leaf versus multiple galls per leaf. Galls widely separated from each other showed less gall larvae mortality due to parasites than galls that were spaced at several galls per leaf. This suggests that parasitic insects conserve energy by concentrating their efforts in small areas rich in opportunities.

In a study of the sawfly *Pontania proxima,* a gall inducer on willows in Ireland and the eastern United States, researchers found that the natural defenses of the larvae against oviposition and attack by parasites, especially the wasp *Pnigalio nemati,* increased with the development of the larvae, particularly in the fourth and fifth **INSTARS**. As the gall-inducer's larvae became more aggressive with age, there was increased failure to complete oviposition once begun by the parasites. When contacted by parasites, sawfly larvae jerk, bite, and strike with their head and the posterior end of their abdomen. As an additional hedge against parasitism, sawfly larvae also continuously clean the interior of the gall, discarding any **FRASS** to the outside, which might be used by parasites as a cue to the presence of larvae.

Of all the *Pontania proxima* galls examined in this study, 23 percent were empty with no signs of any *Pontania* eggs or larvae. Since there was no additional evidence to suggest that the galls had been damaged or opened, it was assumed that the galls had always been empty. Since the galls of *Pontania* spp. begin development as a result of chemicals released by the adult and are well developed by the time the eggs hatch, it was concluded that far more galls were induced than the number of eggs laid. Based on this, it has been suggested that these empty galls may act as decoys, distracting parasites from occupied galls, while increasing the time and distance for the parasites to locate actual hosts. Only 2 percent of the galls of this sawfly were actually parasitized, which may lend credence to this theory.

Finally, other researchers found that parasites had more success in parasitizing the **BISEXUAL**, leaf-galling generation of the cynipid wasp *Callirhytis cornigera* than the **UNISEXUAL (AGAMIC)** generation of stem gallers. The larvae of the latter generation were more protected by the thickness and density of woody tissues than the bisexual generation larvae covered by soft, thin leaf tissues. Such a high rate of survival among the unisexual generation females allowed the population to build over time even though 60 to 80 percent of the bisexual generation would be lost to parasitism.

References: Smith 1970; Washburn and Cornell 1979; Al-Saffar and Aldrich 1998; Eliason and Potter 2001.

Honeydew and Bees, Yellow Jackets, and Ants

The presence of "antagonistic" insects such as honeybees, yellow jackets, and ants prevents inquilines, parasites, and hyperparasites from gaining access to a gall and successfully entering or ovipositing their eggs into the eggs or larvae of the gall inducers. Four separate studies have shown, for example, that ants significantly reduce parasite and inquiline access to galls. The key players in this ecological twist include several gall-inducing cynipid wasps (Flat-topped Honeydew Gall Wasp *[Disholcaspis eldoradensis]*, Twig Gall Wasp *[D. mellifica]*, Round Honeydew Gall Wasp *[D. canescens]*, Clasping Twig Gall Wasp *[D. prehensa]*, Acorn Gall Wasp *[Callirhytis carmelensis]*, and Kidney Stem Gall Wasp *[Andricus reniformis]*), among others. These gall wasps stimulate the production and release of sweet phloem exudates (referred to as honeydew) on the surfaces of their galls during the vulnerable period in the growth of the gall and the development of the larvae. This sweet exudate attracts ants, bees, and yellow jackets that do not tolerate the presence of other species, making

Plate 2.
The galls of *Disholcaspis canescens* secrete a sweet honeydew that attracts ants and yellow jackets.

it nearly impossible for parasitic wasps and others to do their work. Honeydew is known as an important food resource especially for yellow jackets in late summer. One study showed that the yellow jacket *Vespula pensylvanica* was a frequent forager of honeydew produced on coast live oaks by the Acorn Gall Wasp. In one instance a queen yellow jacket visited the same gall four times in two hours to collect honeydew. When a different species of wasp showed up, the queen interrupted her feeding to face the intruder by raising her forebody and front legs in an apparent threat posture, causing the incoming wasp to veer away. In another case involving the same species, a queen yellow jacket returned to a gall previously visited only to find an intruding wasp (different species). The queen attacked the intruder, and the two grappling wasps tumbled off the gall to the ground. I have also seen this behavior involving the Flat-topped Honeydew Gall Wasp on Oregon oaks *(Quercus garryana)*. It seems that once queen yellow jackets start harvesting honeydew, they claim a specific gall as their territory and drive off trespassers, as I have witnessed major battles between queens and other intruding yellow jackets.

In August and September some oak trees are humming with bees and yellow jackets and columns of ants taking advantage of this late-season source of sweet nutrients (pl. 2). I have also seen the galls of the Flat-topped Honeydew Gall Wasp on Oregon oak oozing with honeydew and covered with yellow jackets on one branch, while a nearby branch with the same galls was swarming with ants, indicating an intolerance for each other. Standing back from this rather small tree, I saw hundreds of yellow jackets swarming about the tree as if there were a major hive in it. In the

San Joaquin Valley honeybees gather an estimated 13 to 18 kg (30 to 40 lb) of this sweet honey in a single season, which can actually lead to their early death due to the high mineral content of the exudates.

References: Mani 1964; Felt 1965; Hutchins 1969; Russo 1979; Gambino 1990; Stone et al. 2002.

Insect Predators

The common insect threats to cynipid and tenthredinid gall inducers include chalcid, torymid, braconid, ichneumon, and pteromalid wasps—all parasites and hyperparasites—and beetles and moths as inquilines (sometimes also as predators). Similarly, cecidomyiid larvae are often attacked by wasps from several families. There are some nongalling cecidomyiid midges that prey on gall midge larvae and take over their "home" rather than induce their own. In a rather rare case, the tephritid *Oxyna palpalis* serves as an inquiline in the gall of the midge *Rhopalomyia florella,* on the Great Basin sagebrush *Artemisia tridentata,* in southern California, but becomes a major predator of the midge larvae. In another rare case, a syrphid fly was discovered preying on the **ADELGIDS** (related to aphids) inside a cone gall. Some parasitic mites attack either gall-inducing mites or other gall insects. Also, certain dance flies (Empididae) catch adult gall midges in flight for food and courtship with females. The empidids that eat the gall midges may well, in turn, be captured and eaten by spiders, dragonflies, damselflies, and numerous birds.

This seemingly complex interplay of host, inquiline, parasite, hyperparasite, and predator is best seen in the example of the Willow Apple Gall Sawfly *(Pontania californica).* At least eight different species of insects act as inquilines or parasites in the galls of this sawfly. These include six wasps, a moth, and a weevil. The moth and weevil are inquilines that feed mainly on gall tissue. The moth larvae, however, will kill intruders upon contact, including their own kind. The same is apparently true for the weevil. The adults of parasitic wasps probe with their long ovipositors, which have built-in heat receptors to pick up metabolic heat from the potential hosts they require for their eggs. In some cases, adult parasites sting gall larvae to paralyze them before depositing their eggs. Surviving to complete its life cycle may not be as easy as one might assume for such a tiny insect as the Willow Apple Gall Sawfly.

References: Caltagirone 1964; Mitchell and Maksymov 1977; Sinclair et al. 1987; Gagné 1989; Goeden 2002a.

Birds and Mammals

Another large group of predators represents a serious threat to gall insects at some point during their lives. Fallen leaf galls and gall-inducer **PUPAE** hidden among leaf litter are good sources of protein for shrews, mice, wood rats, and squirrels. Birds such as California Towhees *(Pipilo crissalis)*, Spotted Towhees *(P. maculatus)*, and Fox Sparrows *(Passerella iliaca)* probe leaf litter for seeds and insects. Many galls can be easily mistaken for seeds. While Western Scrub Jays *(Aphelocoma californica)* and Steller's Jays *(Cyanocitta stelleri)* are known to eat small galls, sapsuckers *(Sphyrapicus* spp.) dig into large galls to extract the occupants, as in the case of the California Gall Wasp. During the fall and winter months when these galls are loaded with numerous species of insects and the tissues are spongy and drier than during the growing season, sapsuckers and maybe other woodpeckers can chip out inch-wide holes in search of the tender morsels inside. Other birds including Evening Grosbeaks *(Coccothraustes vespertinus)* have been observed opening aphid roll galls and eating the occupants. Cedar Waxwings *(Bombycilla cedrorum)* have also been seen picking small galls off the undersides of willow leaves. Black-capped Chickadees *(Poecile atricapilla)* and Downy Woodpeckers *(Picoides pubescens)* have been observed digging into goldenrod galls for their larvae. Bushtits *(Psaltriparus minimus)* have been observed pecking into the succulent galls of the Stem Gall Tephritid *(Eutreta diana)* on Great Basin sagebrush.

If adult gall insects hatch and attempt to fly up to their host's branches, they become vulnerable to fly-catching birds (phoebes [*Sayornis* spp.] and others) enroute, and insect-eaters such as vireos (*Vireo* spp.), kinglets (*Regulus* spp.), chickadees (*Poecile* spp.), and nuthatches (*Sitta* spp.), among others, once they land on leaves or branches.

I have found the galls of the Beaked Spindle Gall Wasp on branches of huckleberry oak *(Q. vaccinifolia)* close to the ground that have had their tips chewed off by chipmunks (*Eutamias* spp.) and/or Golden-mantled Ground Squirrels *(Spermophilus lateralis)* to get at the larvae (pl. 3). Similarly, in the same area of the central Sierra, I found the rather large galls of a sawfly, *Blenno-*

Plate 3. The gall of *Heteroecus pacificus* on huckleberry oak chewed open by a Golden-mantled Ground Squirrel or chipmunk to get at the larvae.

Plate 4. The galls of *Blennogeneris spissipes* on creeping snowberry are chewed open by Golden-mantled Ground Squirrels and chipmunks.

generis spissipes, on creeping snowberry (*Symphoricarpos mollis*) that were ravaged by chipmunks and/or Golden-mantled Ground Squirrels (pl. 4). Also, the Western Gray Squirrel (*Sciurus griseus*) chews into the integral stem galls of the Tapered Stem Gall Wasp (*Andricus spectabilis*) to expose the larval chambers and larvae.

Some gall inducers escape this onslaught of dangers and successfully lay eggs to continue their species. As nature would have it, life is not easy, and each species must face the perils of a world with many other creatures looking for food. This incredible interplay of predator and prey seems to never end as one studies the exchange of energy throughout a food chain and food web starting with gall inducers. In nature's ever-fascinating way, the apparent victor (for the moment) may well fall victim to yet another predator higher up the food chain. When you consider that a single species of gall insect and its galls may support a dozen or more species of parasites, hyperparasites, inquilines, and predators of all sorts, you begin to grasp the enormity of the ecological interrelationships that exist just beyond our normal view.

References: Leech 1948; Sullivan 1987; Abrahamson and Weis 1997; Goeden 2002b.

THE CAST OF CHARACTERS in the world of plant galls is quite extensive and worldwide. Plant galls exist on every continent except Antarctica. Even in the Arctic tundra, there are galls on prostrate willows (*Salix* spp.) and huckleberries (*Vaccinium* spp.), among other plant species. The prominence of any one gall-inducing group varies from continent to continent. While cynipid wasps (Cynipidae) dominate North America, for example, mites and midges are the principle gall inducers in Asia.

In this section, the basic biology of all the groups of major gall inducers found in western states is discussed. Individual behaviors and interactions are presented in the gall accounts. The primary gall inducers described herein include bacteria, rust and sac fungi, mistletoes, eriophyid mites, psyllids, aphids, moths, beetles, leaf-mining flies, tephritid fruit flies, gall midges, chalcid wasps, sawflies, and cynipid wasps.

Bacteria

A few galls induced by bacteria are worthy of note in this guide. Most are associated with roots or the crowns of trees. One species, yet to be determined, is associated with eruptions and swellings that occur on the main branches of Douglas-fir *(Pseudotsuga menziesii)*, particularly young trees. Of particular interest is the crown gall bacterium *(Agrobacterium tumefaciens)*, which is associated with large galls on fruit trees and some shrubs, just at the surface of the soil, or on the roots or aboveground parts of the plants as discussed in the "Damage to Host Plants" section in the previous chapter. Another group of bacteria is associated with the production of root nodules on a number of native plants, particularly those that get started in nutrient-poor soil where nitrogen is generally scarce. Leguminous plants such as lupines (*Lupinus* spp.) and alfalfa *(Medicago sativa)* form nodules around invading bacteria of the genus *Rhizobium,* while several nonleguminous plants benefit from their association with bac-teria in the genus *Frankia* (pl. 5). The development of these root nodules varies between the two groups of bacteria, but the end result is that the host plants are able to assimilate atmospheric nitrogen that has been converted by bacterial enzymes into usable ammonium ions. Over 100 species of nonleguminous plants in the United States have *Frankia*-induced nitrogen-fixing

Plate 5. The nitrogen-fixing nodules of *Frankia* sp. on a Sitka alder seedling.

nodules on their roots. This includes over 30 species of *Ceanothus*. Nodules generally form within the upper 30 cm (1 ft) of soil, with some just under the surface. Nodules begin development when a strain of *Frankia* bacteria invades damaged or deformed root hairs. Growth and activity depend on a variety of environmental conditions including soil moisture, temperatures, season, and oxygen levels. In general, nitrogen-fixing plants are able to survive in environments such as outwash plains, flood zones, and landslide and avalanche areas that are missing certain plant nutrients, including nitrogen, and are otherwise unsuitable for species of plants that are more dependent on higher levels of organic material in the soil. By getting established with the help of bacteria in such inhospitable places, these "pioneer plants" not only add available nitrogen to the soil, but they also provide shade and moist microsites for seedlings of other species; they capture seeds blown by wind; and they stabilize the soil. Nitrogen fixation may not be the only element that fosters plant succession in barren environments. The ability of these pioneer plants to establish a foothold in such places helps set the stage for the natural succession of an area into a climax forest over many decades.

References: Sinclair et al. 1987; Walls and Zamora 2001.

Fungi (Families Exobasidiales, Uredinales, and Taphrinales)

Three major groups of fungi act as gall inducers in the West. These include *Exobasidium* fungi (Basidiomycotina, Exobasidiales), rust fungi (Basidiomycotina, Uredinales), and sac fungi (Ascomycotina, Taphrinales). In addition, a powdery mildew, *Cystotheca lanestris* (Ascomycotina, Erysiphales), causes witches' brooms on coast live oak *(Quercus agrifolia)*, interior live oak *(Q. wislizenii)*, canyon live oak *(Q. chrysolepis)*, and valley oak *(Q. lobata)*. These are discussed in the gall accounts.

Unlike plants, fungi lack chlorophyll and cannot produce their own food. As a result, fungi act as either parasites (depending on live tissues) or saphrophytes (depending on dead tissues) for their nourishment. Reproduction is by means of microscopic spores, which are usually released from infected tissues produced on the host plants. Many fungal spores are transported by winds to new hosts. In the absence of a new host, some of these spores can survive for years even at high altitudes (13,000 m [42,640 ft] above sea level). If a viable spore lands on the proper host or substrate, under the right conditions it may germinate, producing fungal threads called hyphae. The hyphae combine to form larger mycelial threads that also act as penetrating strands, drawing nourishment from their hosts. Sometimes, an individual hypha can penetrate host tissues. While most infections are systemic, the presence of these irritating invaders within host plants induces the development of swellings or witches' brooms on trunks and stems, and sometimes swollen leaves or fruit. Generally, these swellings are the sites of spore production. The life cycles of the three major groups of fungi vary greatly from one another, with no one scenario applying to all.

EXOBASIDIUM FUNGI attack false azaleas *(Menziesia ferruginea)*, manzanitas *(Arctostaphylos* spp.), huckleberries, bog rosemary *(Andromeda polifolia)*, crowberry *(Empetrum nigrim)*, rhododendrons and azaleas *(Rhododendron* spp.), and Labrador tea *(Ledum groenlandicum)*. *Exobasidium* gets its name from the habit of producing its spores on the surface of the plant part affected rather than from its own fruiting body, as seen in mushrooms. Some species of *Exobasidium* produce indoleacetic acid, which is thought to play a role in gall formation. Some of these fungi act systemically, while others attack local organs from the

Plate 6. A fungus gall on false azalea caused by *Exobasidium vaccinii*.

Plate 7. A rust gall on shore pine in Alaska caused by *Endocronartium harknessii*.

outside, creating swollen leaves or witches' brooms. The swollen, galled leaves of false azalea usually have a white appearance, reflective of the spores present. Once the spores have spread, the leaf tissues brown and die. The *Exobasidium* galls that appear on manzanita and false azalea leaves look like massive, globular swellings on one side with, corresponding pockets or depressions on the opposite side (pl. 6).

RUST FUNGI have complicated life cycles that often involve more than one species of host plant. Rust fungi go through as many as four different spore-producing stages, yielding four functionally different kinds of spores. Some rusts go through all four stages, while others omit one or more stages, producing fruiting spores after just one stage.

Rust fungi are often not restricted to a species but may attack several species within a genus or use several different alternate seasonal hosts. The western gall rust fungus (*Endocronartium harknessii*) for example, attacks nearly all western pines, creating

TABLE 6 Galls Induced by Fungi and Producing Witches' Brooms or Swellings

Host Plant	Fungus	Type of Gall
Blue spruce (Picea pungens)	Chrysomyxa arctostaphyli	Witches' broom
Coast live oak (Quercus agrifolia)	Live oak witches' broom fungus (Cystotheca lanestris)	Witches' broom, swelling
Cypress		
Alaska-cedar (Cupressus nootkatensis)	Gymnosporangium nootkatense	Swelling
Arizona cypress (C. arizonica)	G. cunninghamianum, Uredo cupressicola	Swelling
Pygmy cypress (C. goveniana subsp. pigmaea)	U. cupressicola	Swelling
Coyote brush (Baccharis pilularis)	Baccharis rust gall fungus (Puccinia evadens)	Witches' broom, swelling
Fir (Abies spp.)	Melampsorella caryophyllacearum	Witches' broom, swelling causing trunk burls
Huckleberry (Vaccinium spp.)	Pucciniastrum goeppertianum	Witches' broom, swelling
Incense-cedar (Calocedrus decurrens)	Incense-cedar rust (G. libocedri)	Witches' broom, swelling
Juniper (Juniperus spp.)	Cedar apple rust (G. juniperi-virginianae)	Swelling (Rocky Mountain area)
	G. aurantiacum	Swelling
	G. confusum	Swelling
	G. inconspicuum	Swelling (Utah)
	G. nidus-avis	Witches' broom, swelling (Rocky Mountain area)
	G. speciosum	Swelling (Utah)
	G. tremelloides	Swelling
	G. tubulatum	Swelling (Rocky Mountain area)
Manzanita (Arctostaphylos spp.)	Exobasidium vaccinii	Swelling on leaf
Mountain ash (Sorbus spp.)	Cronartium comptoniae	Swelling
Oregon crab apple (Malus fusca)	C. coleosporioides	Swelling
Pine		
Chihuahua pine (Pinus leiophylla)	C. conigenum	Swelling on cone
Jeffrey pine (P. jeffreyi)	C. coleosporioides	Swelling

Lodgepole pine (*P. contorta* subsp. *murrayana*)	*C. coleosporioides*	Swelling
Ponderosa pine (*P. ponderosa*)	*C. coleosporioides*	Swelling
Pinus spp.	*C. comandrae*	Swelling
	C. comptoniae	Swelling
	Western gall rust fungus (*Endocronartium harknessii*)	Witches' broom, swelling
Wax myrtle	*Cronartium comptoniae*	Swelling

TABLE 7 Taphrina Galls

Host Plant	Taphrina Species	Type of Gall
Alder (*Alnus* spp.)	Alder cone gall fungus (*Taphrina occidentalis*)	Cone tongues (Pacific states)
	T. amentorum	Cone tongues (Alaska)
	T. robinsoniana	Cone tongues (East Coast)
Buckeye (*Aesculus* spp.)	*T. aesculi*	Broom
Cottonwood (*Populus* spp.)	*T. populi-salicis*	Leaf pockets
Nectarine, Peach (*Prunus* spp.)	Peach leaf-curl fungus (*T. deformans*)	Leaf curl
Plum, Choke-cherry (*Prunus* spp.)	Plum leaf-curl fungus (*T. flectans*)	Leaf curl
	Plum pocket gall fungus (*T. pruni-subcordatae*)	Bladder plums
	Witches' broom fungus (*T. confusa*)	Broom
Service-berry (*Amelanchier* spp.)	Service-berry fungus (*T. amelanchieri*)	Broom
Wood fern (*Dryopteris arguta*)	Wood fern sac fungus (*T. californica*)	Leaf gall

Note: Cone tongues are enlarged bracts.

large knobby galls on branches and roots (pl. 7). Incense-cedar rust *(Gymnosporangium libocedri)* alternates between incense-cedar *(Calocedrus decurrens)* and service-berry (*Amelanchier* spp.).

Rusts often initiate integral stem swellings from which orange powdery spores are released. Some rust fungi appear as simple orange, swollen blisters on leaves of service-berries, grasses, and roses (*Rosa* spp.), among others. Still other rust fungi produce long, gelatinous brown or orange fingers or slimy blobs that hang from needles, galls, and branches, as in cedar apple rust *(G. juniperi-virginianae)* on Rocky Mountain juniper *(Juniperus communis)*. In other examples, rust fungi cause witches' brooms that usually die after the growing season, as seen with incense-cedar and incense-cedar rust (see incense-cedar rust account). A number of rust fungi initiate globose swellings of stems and trunks and even develop on exposed large roots. Table 6 lists common plants that host rust-caused galls of the swelling and witches' broom types.

In some rusts, the availability of spores attracts a broad variety of spore-eating insects, as well as other insects that explore the fissures of the galls. One blister rust gall on lodgepole pine *(Pinus contorta* subsp. *murrayana)* was found to support 137 species of insects.

SAC FUNGI are among the most interesting gall-inducing fungi. The spores of sac fungi are produced in microscopic sacs or envelopes and are forcibly discharged in late spring to mid-summer to infect new host tissue. Among the sac fungi, several species of the genus *Taphrina* are widely known for initiating leaf blisters, leaf curls, swollen fruit called **BLADDER PLUMS,** and witches' brooms (table 7). The best known of the 100 or so species in the genus is peach leaf-curl fungus *(T. deformans),* which causes peach leaf curl, a major concern for nursery workers, farmers, and landscapers because it attacks peaches and nectarines. Other species of *Taphrina* attack cottonwood (*Populus* spp.), alder (*Alnus* spp.), service-berry, choke-cherry and wild plum (*Prunus* spp.), and wood fern *(Dryopteris arguta).* A related sac fungus, the black knot gall fungus *(Apiosporina morbosa),* attacks the stems of all members of the genus *Prunus,* initiating large, black knot galls on the stems (see the black knot gall fungus account).

References: Mix 1949; Arthur 1962; Felt 1965; Hepting 1971; Westcott 1971; Sinclair et al. 1987.

Mistletoes (Order Santalales, Family Viscaceae)

Only two genera of mistletoes in California and other western states are prominent inducers of stem swellings or witches' brooms (table 8). While simple, localized stem swellings occur in response to the invasion of mistletoe and are generally not considered true galls, I include here only those species that induce witches' brooms, which I consider to be true galls. These include the genera *Phoradendron* and *Arceuthobium*. A third genus of mistletoe involves a European species introduced to California, *Viscum album,* which is not treated here.

All mistletoes are parasitic on the branches of their hosts. *Arceuthobium* spp. derive water, nutrients, and small quantities of organic compounds from their host trees. In the case of *Phoradendron* spp., however, they obtain most of their nutrients through photosynthesis and gain little from their host. *Phoradendron* spp. have male and female flowers on separate plants. Initial infection begins when a single, viscous seed lands on a tree branch. Once the plant grows and flowers, multiple infections occur through the seeds produced by the original mistletoe. The stems of mistletoe contain chlorophyll, and this is where most photosynthesis takes place even though the plant may have small leaves. Mistletoes in California and the West are usually associated with pine (*Pinus* spp.) (pl. 8), juniper (*Juniperus* spp.), fir (*Abies* spp.), larch (*Larix* spp.), hemlock (*Tsuga* spp.), and desert shrubs such as catclaw *(Acacia greggii).*

References: Sinclair et al. 1987; Scharpf 1993.

Plate 8. An integral stem swelling on pine caused by the dwarf mistletoe *Arceuthobium americanum.*

TABLE 8 Mistletoe Galls

Host Plant	Mistletoe Species	Type of Gall
Catclaw (*Acacia greggii*)	*Phoradendron californicum*	Broom, stem swelling
Cypress (*Cupressus* spp.)	*P. bolleanum*	Stem swelling
Douglas-fir (*Pseudotsuga menziesii*)	Douglas-fir Dwarf Mistletoe (*Arceuthobium douglasii*)	Broom
Fir (*Abies* spp.)	*A. abietinum*	Broom, stem swelling
Hemlock (*Tsuga* spp.)	*A. tsugense*	Broom, stem swelling
Juniper-cypress (*Cupressus* spp.)	*P. bolleanum*	Stem swelling
Juniper-incense-cedar (*Calocedrus decurrens*)	*P. juniperinum*	Stem swelling
Larch (*Larix occidentalis* spp.)	*A. laricis*	Broom, stem swelling
Pine		
Lodgepole pine (*Pinus contorta* subsp. *murrayana*)	*A. americanum* *A. cyanocarpum*	Broom, stem swelling Broom, stem swelling
Pinyon pine (*P. monophylla, P. edulis*)	*A. divaricatum*	Slight stem swelling
Ponderosa pine (*P. ponderosa*)	*A. vaginatum*	Broom
Sugar pine (*P. lambertiana*)	*A. californicum*	Broom, stem swelling
White pine (Arizona) (*P. strobus*)	*A. apachecum*	Broom
Western white pine (*P. monticola*)	*A. cyanocarpum*	Broom, stem swelling
Bishop pine (*P. muricata*)	*A. occidentale*	Broom, stem swelling
Coulter pine (*P. coulteri*)	*A. occidentale*	Broom, stem swelling
Gray pine (*P. sabiniana*)	*A. occidentale*	Broom, stem swelling
Monterey pine (*P. radiata*)	*A. occidentale*	Broom, stem swelling
Coulter pine	*A. campylopodum*	Broom, stem swelling
Jeffrey pine (*P. jeffreyi*)	*A. campylopodum*	Broom, stem swelling
Knobcone pine (*P. attenuata*)	*A. campylopodum*	Broom, stem swelling
Ponderosa pine	*A. campylopodum*	Broom, stem swelling

Note: Most of the mistletoes in the western states cause witches' brooms along with moderate stem swelling.

Mites (Order Acarina, Family Eriophyidae)

Mites are fascinating animals distantly related to spiders and scorpions. For many years they have been recognized as serious pests of livestock, vegetables, and fruit crops. Over the millennia of their evolution, mites have developed ecological niches that would surprise most people—from the fruit of a tree, to the feather of an eagle, to the skin of a pig. Wow!

Some of these mites, called eriophyids, have developed intimate relationships with plants that result in the creation of galls (table 9). While their tiny size (less than .3 mm long) (fig. 11) conceals them from view, the damage that results from their feeding habits is often quite noticeable. The galls that form can look like blisters, globular beads, pouches, clubs, swollen and folded leaf edges, witches' brooms, or

Figure 11. The eriophyid mite *Phytoptus emarginatae* measures .3 mm long.

hair-covered depressions called **ERINEA**. Erineum galls are common on oaks, alders, crabapples (*Malus* spp.), and walnuts (*Juglans* spp.). The bead, pouch, and club galls created by eriophyids are often covered with minute hairs and almost always have openings at the base to allow escape of adults and young. In most cases, eriophyid mites are highly specialized plant-cell feeders that have evolved relationships with specific host plants. In a few cases, eriophyid mites will be found on several species of a genus such as *Juglans* (pl. 9) or *Populus*.

Plate 9. Large bead galls on walnut caused by *Aceria brachytarsus*.

TABLE 9　Eriophyid Mite Galls

Host Plant	Mite Species	Type of Gall
Alder (*Alnus* spp.)	*Acalitus brevitarsus*	Erineum
	Phytoptus laevis	Leaf bead galls
Alfalfa (*Medicago sativa*)	*Eriophyes medicaginis*	Witches' broom
Ash (*Fraxinus* spp.)	*E. fraxinivorus*	Inflorescence gall
Aspen (*Populus* spp.)	*E. neoessigi*	Catkin galls
	Phyllocoptes didelphis	Erineum
Buttonbush (*Cephalanthus* spp.)	*E. cephalanthi*	Leaf bead galls
Chamise (*Adenostoma fasciculatum*)	*Phytoptus adenostomae*	Leaf bead galls
Cottonwood (*Populus* spp.)	*E. neoessigi*	Catkin galls
	E. parapopuli	Bud galls
Coyote brush (*Baccharis pilularis*)	*E. baccharipha*	Leaf pit galls
Douglas-fir (*Pseudotsuga menziesii*)	*Tricetacus pseudotsugae*	Stunted needles
Gooseberry (*Ribes* spp.)	*E. breakeyi*	Leaf bead galls
Grape (*Vitis* spp.)	*Colomerus vitis*	Erineum
Huckleberry (*Vaccinium* spp.)	*Acalitus vaccinii*	Bud/berry galls
Juniper (*Juniperus* spp.)	*T. quadricetus*	Deformed berry
Meadowsweet (*Spiraea* spp.)	*Phytoptus paraspiraeae*	Flower galls
Mountain maple (*Acer glabrum*)	*E. calaceris*	Erineum
Oak		
Live oak (*Quercus* spp.)	*E. paramackiei*	Witches' brooms
	E. mackiei	Erineum
Canyon live oak (*Q. chrysolepis*)	*E. mackiei*	Erineum
Blue oak (*Q. douglasii*)	*E. trichophila*	Erineum
Pine (*Pinus* spp.)	*T. alborum*	Stunted needles
Plum (*Prunus* spp.)	*Phytoptus emarginatae*	Leaf nail galls
Poison oak (*Toxicodendron diversilobum*)	*Aculops toxicophagus*	Leaf bead galls
Sagebrush (*Artemisia* spp.)	*Aceria paracalifornicus*	Leaf pit galls
Scale broom (*Lepidospartum squamatum*)	*E. lepidosparti*	Bud galls
Seep-willow (*Baccharis salicifolia*)	*E. baccharices*	Leaf pit galls Wartlike galls
Skeletonweed (*Chondrilla juncea*)	*E. chondrillae*	Bud galls
Snowberry (*Symphoricarpos* spp.)	*Phyllocoptes triacis*	Leaf roll galls

Tobacco brush (Ceanothus velutinus)	E. ceanothi	Leaf bead galls
Walnuts		
California black walnut (Juglans californica)	E. caulis	Petiole galls
California black walnut (J. californica)	E. brachytarsus E. neobeevori	Leaf bead galls Catkin gall
English walnut (J. regia)	E. erineus	Erineum
Hind's walnut (J. hindsii)	E. brachytarsus	Leaf bead galls
	E. spermaphaga	Catkin galls
Willow (Salix spp.)	Aculops aenigma	Bud/catkin galls
	A. tetanothrix	Leaf bead galls

Note: This list includes many of the eriophyid mite galls known to occur in the western states.

Gall formation occurs as a result of the attack on individual plant cells. With their piercing mouthparts, eriophyid mites are able to drain the contents of plant cells. A single egg-laden female can initiate gall formation through piercing and feeding on plant cells while releasing chemicals into the wounds during the process. Females who do not come into contact with the **SPERMATOPHORES** (sacs containing sperm) left by males will, in turn, produce only males. Females that do contact male spermatophores will produce males and females. In some cases, females retain sperm through the winter, producing their young in the spring. Offspring remain in the gall with their parent, contributing to further gall development through their feeding activities. Young mites pass through two nymph stages, with the adults developing from the second stage. Growth from egg to adult often takes only two weeks. Adult females often hibernate in fissures in the bark or among bud scales during winter. Males generally do not survive winter. Mites travel by drifting in the wind and riding on insects and birds that come into contact with host plants. This hitchhiking helps dispersion to new host plants. The galls initiated by eriophyid mites are usually much more difficult to discern as mite galls than the more distinct galls caused by other organisms.

References: Keifer 1952; Oldfield et al. 1970; Keifer et al. 1982; Johnson and Lyon 1991.

Aphids and Adelgids

(Order Hemiptera, Families Aphididae and Adelgidae)

Of the two families represented here, aphids induce most of the galls described in this guide. As a result, the following discussion focuses on aphid biology, noting there are significant differences between the two groups.

Aphids belong to the same order of insects as leafhoppers, spittlebugs, mealybugs, and scale insects. In one way or another, aphids have long been considered serious pests by nursery workers and home gardeners. Galling species of aphids represent a small proportion (approximately 10 percent) of all 4,400 aphid species known. They feed by piercing the epidermis of a leaf or bud with their long, needlelike mouthparts, called **STYLETS**, and sucking the juice from the phloem tissue. Aphids have the ability to convert plant starch into soluble sugars through enzymes in their saliva. Several studies have concluded that aphid saliva contains indole-3-acetic acid and that this auxin is the principal gall-stimulating agent in aphid saliva. The shapes of aphid galls are species specific.

Aphids have complex life cycles, particularly among gall-inducing species. Most apparently have single hosts, for example, *Tamalia* spp. Others, such as *Pemphigus* spp., have alternate hosts. Many species also engage in an alternation of generations, as do some cynipid wasps.

For the most part, gall aphids go through several parthenogenetic generations, culminating in a sexual generation that lays its fertile eggs on the bark of primary host trees in fall. Here, the eggs overwinter. In spring, an adult aphid called a **FUNDATRIX**, or **STEM MOTHER**, emerges, initiates a gall, and reproduces parthenogenetically. With some species, stem mothers display strong territorial behavior and will engage in intraspecific fights for prime galling sites. The fundatrix and later generations of wingless females are capable of initiating gall formation. One interesting aspect of specialization in galling aphids is the development of sterile morphs called soldiers, who apparently have a defensive function against potential predators. They react to colony disturbance and will bite or cling to intruders. In some cases, soldiers remove excess honeydew and stand guard at the entrance to the galls. The rest of the stem mother's offspring develop wings

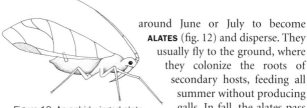

Figure 12. An aphid winged alate.

around June or July to become **ALATES** (fig. 12) and disperse. They usually fly to the ground, where they colonize the roots of secondary hosts, feeding all summer without producing galls. In fall, the alates pass into another stage, and they fly to the primary host tree or shrub. These individuals produce the males and females that overwinter, hatching into stem mothers the following spring. Two separate groups of aphidlike insects and their galls are considered here. One group comprises the true aphids, belonging to the family Aphididae (*Pemphigus* spp. and *Tamalia* spp.) (pl. 10), while the second group has been mistak-

Plate 10.
A petiole gall on black cottonwood caused by *Pemphigus populicaulis*.

enly called aphids when their members actually belong to the family Adelgidae (*Adelges* spp.). The galls of *Pemphigus* spp. typically occur on cottonwood petioles or leaves. The red pink galls of the Manzanita Leaf Gall Aphid *(Tamalia coweni)* occur on the new leaves of nearly all **GLABROUS** manzanitas. One stage of this aphid also galls manzanita inflorescences.

Fourteen adelgid species are associated with conifers in the United States. Of these, *Adelges cooleyi* and *A. piceae* are among the most prominent. The galls of *A. cooleyi* occur on new terminal twigs of spruce (*Picea* spp.) as do those of *A. piceae* on fir. More details on these species are presented in the gall accounts.

References: Palmer 1952; Harper 1959, 1966; Lange 1965; Grigarick and Lange 1968; Alleyne and Morrison 1977; Johnson and Lyon 1991; Moran 1992; Shorthouse and Rohfritsch 1992; Miller 1998, 2004; Miller and Sharkey 2000; Cranshaw 2004; Wool 2004.

Psyllids (Order Hemiptera, Family Psyllidae)

Psyllids (fig. 13), or jumping plant lice, are important gall inducers all over the world, while in the West they have little or no role in galling native plant species. They are much less common in North America than other gall inducers, with about 11 species known. Of all the gall-inducing psyllids known worldwide (approximately 350 species), nearly half are from Asia and Africa. Gall-inducing psyllids usually initiate pit or pouch galls on leaves, although in some areas they induce leaf-roll galls. These galls differ from those of erio-

Figure 13. A generalized psyllid.

phyid mites in that they lack the erineum (hair), in addition to having thicker walls and more regular shapes. The newly hatched nymph of the psyllid is the principal in initiating gall formation. Offspring of the adults can occur in the galls of the parents. Like other insects, psyllids suck cellular sap with their piercing mouthparts. The psyllid that galls hackberry (*Celtus* spp.) inserts its stylets into the host leaf and stimulates a "**CELLULO-SOLYTIC EFFECT**," causing dissolution of the intercellular walls. Giant multinucleate cells are formed as a result. Cells begin differentiating beneath these giant cells, which ultimately leads to the formation of the galls. In the West, psyllids gall the ornamental Australian brush-cherry *(Eugenia myrtifolia)* (pl. 335), ornamental pepper tree *(Schinus molle)* (pl. 11), and hackberry *(Celtis reticulata)*. Also, a nongalling lerp psyllid occurs on the leaves of blue gum eucalyptus *(Eucalyptus globulus)*.

References: Mani 1964; Meyer 1987; Cranshaw 2004.

Plate 11. Pit galls on the leaflets of pepper tree caused by the psyllid *Calophya rubra*.

Moths (Order Lepidoptera, Families Gelechiidae, Tortricidae, and Cosmopterigidae)

Several gall-inducing moths are treated in this field guide; however, numerous others are known to occur in the West (see table 10 for more species). Worldwide, 352 gall-inducing lepidopterans are currently known. The family Gelechiidae has the largest number of gall inducers, with 47 species in 20 genera. Povolny (2003) described 25 species of *Gnorimoschema* occuring in the West. The genera covered here include *Gnorimoschema* (Gelechiidae) on coyote brush *(Baccharis pilularis)*, desert broom *(B. sarothroides)*, rabbitbrush *(Chrysothamnus* spp.), and gumplant *(Grin-*

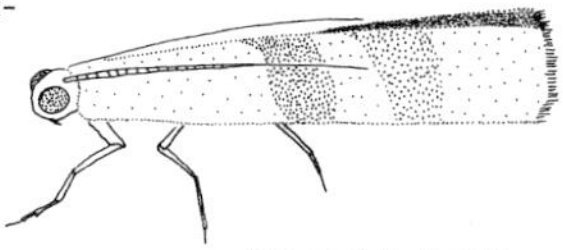

Figure 14. A gall-inducing moth similar to *Eugnosta* sp.

delia hirsutula); *Scrobipalpopsis* (Gelechiidae) on horsebrush (*Tetradymia* spp.); *Sorhagenia* and *Periploca* (Cosmopterigidae) on coffeeberry (*Rhamnus* spp.) and ceanothus (*Ceanothus* spp.), respectively; and *Eugnosta* (Tortricidae) (fig. 14) on cheesebush *(Hymenoclea salsola)*. Please keep in mind that there are far more gall-inducing moths out there than presented in this guide. In addition, several species are inquilines in the galls of other insects or pathogens. The inquiline *Batrachedra salicipomonella* occurs in *Pontania* galls on willows, and while largely herbivorous, it will prey on any insect encountered within the galls. Another example is the inquiline *Dioryctria banksiella,* which feeds on the galled tissues of jack pine *(Pinus banksiana)* but not the *Endocronartium harknessii* that induces the galls. The ecology of gall-inducing insects and their inquilines and parasites is vastly complicated and ripe with research opportunities.

While each species has significant differences from the others, the basic biology of the Baccharis Stem Gall Moth *(Gnorimoschema baccharisella)* on coyote brush (pl. 12) makes a good sample life history. The eggs of this moth are deposited on peripheral branches in fall, where they overwinter. Newly hatched larvae seek new growing shoots and burrow in sometime after the first of the year. The gall begins growing at the point of entry and surrounds the larvae. These spindle-shaped swellings are evident by February. By June or July, many larvae have been lost to parasites. At maturity, usually June to August, the surviving

TABLE 10 Moth Galls

Moth Family and Species	Host Plant	Type of Gall
Argyresthiidae		
Argyresthia pseudotsuga	Douglas-fir *(Pseudotsuga menziesii)*	Terminal twig swelling
Cosmopterigidae		
Ceanothus Stem Gall Moth *(Periploca ceanothiella)*	Ceanothus *(Ceanothus* spp.)	Tapered stem gall
Midrib Gall Moth *(Sorhagenia nimbosa)*	Coffeeberry *(Rhamnus* spp.)	Leaf swelling
Stagmatophora enchrysa	Bluecurls *(Trichostema* sp.)	Stem gall
S. iridella	Bluecurls *(Trichostema* sp.)	Root gall
Gelechiidae		
Baccharis Stem Gall Moth *(Gnorimoschema baccharisella)*	Coyote brush *(Baccharis pilularis)*	Tapered stem gall
Bent Gall Moth *(G.* cf. *octomaculellum)*	Great Basin sagebrush *(Artemisia tridentata)*	Tip gall
Bent Gall Moth *(G. octomaculellum)*	Rabbitbrush *(Chrysothamnus* spp.)	Terminal leaf gall
Eurysaccoides gallaespinosae	Horsebrush *(Tetradymia* sp.)	Stem gall
G. coquillettella	Goldenbush *(Ericameria* spp.)	Onion-shaped stem galls
G. crypticum	*Isocoma menziesii*	Soft, fleshy stem gall
G. crypticum	Sawtoothed goldenbush *(Hazardia squarrosa)*	Soft, fleshy stem galls
G. ericameriae	Goldenbush *(Ericameria* spp.)	Terminal bud gall
G. subterraneum	*Aster chilensis*	Watermelon-shaped stem galls
Gnorimoschema sp.	Desert Broom *(Baccharis sarothroides)*	Terminal leaf gall
Gumweed Gall Moth *(G. grindeliae)*	Gumplant *(Grindelia hirsutula)*	Soft, elliptical stem galls
Horsebrush Stem Gall Moth *(Scrobipalpopsis tetradymiella)*	Horsebrush *(Tetradymia stenolepis)*	Tapered stem gall
Scrobipalpopsis sp.	Scale broom *(Lepidospartum squamatum)*	Stem gall
Momphidae		
Mompha sp.	California fuchsia *(Epilobium canum)*	Stem gall
Nepticulidae		
Ectoedemia sp.	*Populus* sp.	Petiole gall

Plutellidae		
Ypsolopha sp.	*Ephedra* sp.	Stem gall
Tortricidae		
Epiblema rudei	Snakeweed (*Gutierrezia* spp.)	Stem gall
Eugnosta sp.	Cheesebush (*Hymenoclea salsola*)	Tapered stem gall
Unknown	Cooper's box thorn (*Lycium cooperi*)	Stem gall
Unknown	*Vaccinium membranaceum*	Leaf roll gall

Note: This list includes some of the moth galls known in the West.

Plate 12. Integral stem gall on coyote brush caused by *Gnorimoschema baccharisella.*

larvae cut holes through the gall walls and drop to the ground on silken threads. Pupation occurs fairly soon after emergence, without any diapause. In other areas and with other species, the larvae may form cocoons in leaf litter or sand and spend the winter there. With some of these species, the galls they induce are often close to the ground.

References: Powell 1975; Furniss and Carolin 1977; Johnson and Lyon 1991; Miller 2000, 2005; Powell and Povolny 2001; Povolny 2003.

Beetles (Order Coleoptera, Family Cerambycidae)

Only one beetle is treated here as a gall-inducing insect: *Saperda populnea*, a roundheaded woodborer (fig. 15), and it occurs on cottonwoods. This beetle creates an integral stem gall. Sometime in spring, depending on elevation and weather, adult beetles emerge from stem galls. Fairly soon thereafter, they deposit eggs in the bark of twigs and saplings. Adults chew horseshoe-shaped incisions in the bark of small branches on saplings or near the tips of branches on older trees, where they place a single egg, or sometimes two, in each incision. These dark brown, half-moon incision scars remain visible for years (pl. 13). After hatching, the larvae feed on phloem tissue, mining their way down through the stem. The tree reacts by producing callus tissue around the wound, which accumulates, forming a globose gall two to three times the diameter of the host stem. The entry wounds created by adults open the tree to the invasion of the dangerous pathogenic canker fungus *Hypoxylon mammatum*. Adults complete their life cycle in the galls, emerging the following spring.

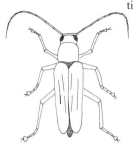

Figure 15. A gall-inducing longhorn wood-boring beetle similar to *Saperda* sp.

References: Johnson and Lyon 1991.

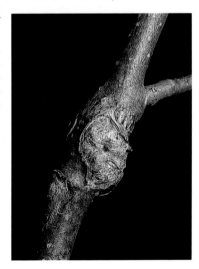

Plate 13. A gall on black cottonwood caused by *Saperda populnea*.

Leaf-mining Flies
(Order Diptera, Family Agromyzidae)

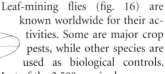

Leaf-mining flies (fig. 16) are known worldwide for their activities. Some are major crop pests, while other species are used as biological controls. Most of the 2,500 species known are leaf- or plant-tissue miners. Larvae typically create erratic trails (mines) in leaves, as shown by the brown epidermal tissue at the surface, or they create "shot holes," where tissue was mined in circular patterns. For the most part, these "mining" flies lay their eggs in plant tissue, where larvae either mine channels through leaf tissue or feed inside roots, stems, or seeds. In the West, a few species induce galls, and two species are prominent. One fly, Poplar Twiggall Fly

Figure 16. A generalized leaf-mining fly.

(Hexomyza schineri), creates large integral stem galls on aspens *(Populus tremuloides)* (pl. 14) and some cottonwoods; the other fly, *Ophiomyia atriplicis,* creates leafy bud galls on saltbush (*Atriplex* spp.). These gall inducers usually overwinter as larvae and either pupate inside the galls or drop to the ground, where they pupate in the soil.

References: Gagné 1989; Johnson and Lyon 1991; McGavin 2002; Cranshaw 2004.

Plate 14. The galls of *Hexomyza schineri,* a leaf-mining fly, on aspen saplings.

Tephritid Fruit Flies

(Order Diptera, Family Tephritidae)

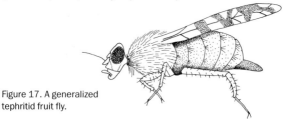

Figure 17. A generalized tephritid fruit fly.

Our knowledge of the role of tephritid fruit flies (fig. 17) as gall-inducing agents has become much clearer over the last 20 years, based largely on the works of Richard Goeden, James Wangberg, and David Headrick, among others. This fly family is one of the largest and, perhaps, most famous, with about 4,200 species known worldwide. While many members of the family are **FRUGIVOROUS** (feed on fruit), many others are considered **NONFRUGIVOROUS**, and it is among these latter flies that we find the gall inducers. With a few exceptions, most gall-inducing tephritids rely on plants in the sunflower family (table 11). These flies initiate galls on both aerial and subterranean plant tissues. The adults emerge at a time when the host plants are suitable for oviposition; they reproduce sexually, and deposit eggs in terminal or axillary buds. A few species deposit eggs in immature flower heads. Courtship and mating are highly complicated and extremely varied among the species. Adults live for 25 to 30 days and can lay up to 150 eggs in different locations, either singly or in clutches of up to 16. Some species can overwinter as adults, living up to 200 days. It is not unusual for larval stages to last for some time, as the galls they inhabit develop over a long period of time, sometimes from several months to nearly a year. It appears that some species even overwinter in a third-instar larval stage. Some species are **FREEZE TOLERANT**, producing **CRYOPROTECTANT** compounds such as sorbitol and glycerol to slowly buffer internal gall temperatures from freezing conditions on the outside. This cold hardiness is especially important for the numerous species found on high-altitude plants such as Great Basin sagebrush (*Artemisia tridentata*) and rabbitbrush (pl. 15), which are subject to heavy snow cover in winter.

TABLE 11 Tephritid Galls

Host Plant	Tephritid Species	Type of Gall
Burro-weed (Ambrosia dumosa)	Bud Gall Tephritid (Procecidochares kristineae)	Woolly bud with bracts
	P. stonei	Bud gall with bracts
California goldenrod (Solidago californica)	P. anthracina	Terminal or lateral bud galls
Cudweed (Gnaphalium luteo-album)	Trupanea signata	Woolly terminal bud
Desert Broom (Baccharis sarothroides)	Desert Broom Gallfly (Aciurina thoracica)	Integral stem gall
Goldenbush (Isocoma acradenia)	P. blanci	Terminal or lateral bud with bracts
Great Basin sagebrush (Artemisia tridentata)	Oxyna aterrima	Woolly bud gall
	Stem Gall Tephritid (Eutreta diana)	Leafy integral stem
Groundsel (Senecio douglasii)	T. stigmatica	Integral stem gall
Rabbitbush (Chrysothamnus spp.)	A. semilucida	Terminal or lateral bud gall
	Beaked Gall Tephritid (A. michaeli)	Pear-shaped, beaked bud
	Bubble Gall Tephritid (A. trixa)	Bud gall
	Cotton Gall Tephritid (A. bigeloviae)	Woolly bud gall
	Elliptical Stem Gall Tephritid (Valentibulla californica)	Integral stem gall
	Hairy Bud Gall Tephritid (Procecidochares sp. B)	Leafy bud gall
	Leafy Bud Gall Tephritid (Procecidochares sp. A)	Leafy bud gall
	Leafy Cone Gall Tephritid (A. idahoensis)	Bud gall with bracts
	Medusa Gall Tephritid (A. ferruginea)	Bud rosette with bracts
Seep-willow (B. salicifolia)	Tephritis baccharis	Integral stem gall
Silver wormwood (A. ludoviciana)	E. simplex	Leafy bud and stem
Trixis californica	Trupanea conjuncta	Integral stem gall
Woolly bur-sage (Ambrosia eriocentra)	Woolly Flower Gall Tephritid (P. lisae)	Woolly bud with bracts

Note: This list includes some of the known tephritid galls in southern California.

Plate 15. Bud galls on rabbitbrush caused by *Procecidochares* sp. A.

References: Goeden 1987, 1988, 1990, 2002a,b,c; Goeden and Headrick 1991; Headrick and Goeden 1993, 1997, 1998; Goeden et al. 1995; Goeden and Teerink 1996a,b,c, 1997a,b,c; Headrick et al. 1997; Goeden and Norrbom 2001.

Midges (Order Diptera, Family Cecidomyiidae)

The world has over 5,450 species of cecidomyiids (fig. 18), with many others yet to be described. Over 1,150 species are known from North America north of Mexico, and many other species have been recorded but not yet identified. Some cecidomyiids specialize in feeding on fungi or as inquilines in galls of other midges and are not involved in gall-inducing behavior. Of the total known to occur in North America, more than 800 species are associated with galls. Some genera are particularly well represented as gall inducers in North America *(Asphondylia, Rabdophaga, Rhopalomyia,* and *Contarinia).* As with the tephritid fruit flies, the biology of cecidomyiids is varied and complex. Eggs hatch within a few days after deposition. Some eggs are laid directly in plant tissue, while other eggs are simply deposited on the

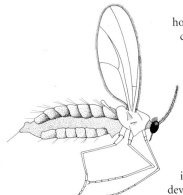

Figure 18. A gall midge, *Asphondylia* sp.

host plant. In the latter case, larvae crawl about until they locate the proper plant organ for gall formation. Many die en route. Galls develop in response to larval secretions and behavior. Larvae that live in leaf-roll galls are usually gregarious, with several within the gall chamber. More complex galls, however, tend to have individual larval chambers. Gall development begins after the larvae commence feeding on plant tissues. Unlike cynipids, when cecidomyiid larvae are feeding on the nutritive cells lining their chambers, they do not destroy the cells because they do not have the chewing mouthparts that cynipids possess. The only result of their sucking is that mild necroses occur. The nutritive cells of cecidomyiid larval chambers have a high metabolic rate, which, like those of cynipids, is controlled by the larvae.

Some cecidomyiids do not initiate their own galls but instead deposit their eggs next to those of gall inducers. The "hitchhiking" midges stay close to the gall-inducing larvae, and once the galls have developed, they act as inquilines and occasional predators.

Upon completion of larval development, some leave the galls to pupate (typical of roll-gall species), while others pupate within the gall, leaving later as adults. Those that leave as larvae drop to the soil, dig in, and pupate. Those that stay with the galls use one of two exit strategies. In one scenario, full-grown larvae cut tunnels through the gall tissue almost to the outside with a specialized, elongate epidermal structure called a spatula, leaving a thin skin at the surface of the gall through which the pupae later escape. In the other case, pupae cut their way through the gall tissues. In both scenarios the mature pupae exit only about two-thirds of the way out, using the pressure of the last part of the passage to hold them firmly in place while they complete their transformation into adults. Emergence of adults is generally well timed to the host plant's biology and growth patterns. Most gall midges are restricted to a species or, in some cases, to related

Plate 16. Hairy bud galls on saltbush caused by *Asphondylia floccosa*.

species within a genus. In a few cases, generalist gall midges feed on members of several plant families (some *Prodiplosis* spp.).

One of the most interesting aspects of cecidomyiid biology is the regular occurrence of fuzzy, white fungi in the galls of all *Asphondylia* (pl. 16) and some *Lasioptera, Kiefferia,* and *Schizomyia* species. These galls are called **AMBROSIA** galls. Gall midges of the *Asteromyia* genus have also been associated with ambrosia galls. The fungi coat the interior walls of the gall, creating a smooth surface. Ambrosia galls tend not to be lined with nutritive cells. Instead, the gall larvae extract food directly from the vascular bundles of their host galls or from the fungal mycelium. In many cases it appears that the fungi serve as food for the larvae (*Asphondylia* spp.), while in other cases the fungus is not eaten, yet completely surrounds the larvae (*Asteromyia* spp.). In this latter situation, the fungus eventually turns black and forms a hardened shell around the larvae, which possibly protects the larvae from parasites. For those species' larvae that feed on the ambrosia fungi, fungus growth appears to be affected, as proposed in the case of the gall midge *Schizomyia galiorum*. After larvae left the galls to pupate in the soil, gall fungi grew, filling the larval chambers. Shorthouse and Rohfritsch (1992, 136) reported, "It appears that the larvae of the four ambrosia gall species take food from fungal mycelium and that they are able to inhibit excessive fungal growth." Much more work needs to be done in this area before these relationships can be proven.

The method by which the fungus is introduced along with

eggs into plant tissue is not definitely known. It is possible the consumed fungal material passes into the pupae and then the adult. In some cases, infected (spore-laden) feces and other secretions are deposited during egg laying, which may allow fungal growth in the gall cavity. In other cases, fungal material may be scooped up by a shoveling action of the rear end of the abdomen of emerging females and subsequently deposited with eggs and accessory secretions. Once the plant cells begin differentiating to form the gall, the fungal spores germinate and rapidly develop mycelia. Nevertheless, the relationship between these symbiotic fungi and their cecidomyiid associates is intriguing and worthy of much greater study.

Finally, I decided not to include a list of cecidomyiids, as I did for other groups such as eriophyids and tephritids, simply because such a list far exceeds the abilities of a small guide.

References: Mani 1964; Gagné 1973, 1975, 1986, 1989, 1995, 2004; Gagné and Hawkins 1983; Gagné and Waring 1990; Shorthouse and Rohfritsch 1992; Gagné and Strong 1993.

Wasps (Order Hymenoptera, Families Chalcididae, Tenthredinidae and Cynipidae)

Wasps in the families Chalcididae, Pteromalidae, Eurytomidae, Agaonidae, Tenthredinidae, and Cynipidae are known to induce galls in North America. Of these several families of gall-inducing wasps, only one family, Cynipidae, is prominent. Only two species of chalcid gall inducers are included here. The tenthredinids, or sawflies, induce significantly fewer galls than do cynipids, but nonetheless warrant mention. Cynipid wasps, by far, are the star gall inducers throughout the United States, especially in the West. As a result of the diversity of habitats in California and, therefore, the high number of oak species (approximately 24), California supports more cynipid gall wasp species than any other state, and more than the entire western Palearctic region.

Unlike the larger, black and yellow wasps that seem to specialize in plaguing barbeque parties and camping trips, cynipid wasps are diminutive insects whose life habits are neither threatening nor dangerous. Actually, of all the gall insects, the cynipids are among the most intriguing in this ecological drama. These wasps have the most complex life cycles of all gall insects, and their galls are structurally complex and tend to be the most colorful.

Chalcid Wasps (Order Hymenoptera, Family Chalcididae)

The western states have several hundred described species of chalcid wasps (fig. 19, pl. 17), and probably many more have yet to be described. One of the more notable chalcids is the symbiotic Fig Wasp *(Blastophaga psenes),* which is critical to the production of figs. Most species are small, metallic black, dark brown, or red and function as parasites or hyperparasites on other

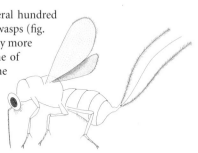

Figure 19. A chalcid wasp.

insects, especially caterpillars and fly larvae. Some are plant feeders whose larvae develop in seeds, while others are gall inducers with specific host-plant associations. Compared with the larger groups of cecidomyiids and cynipids, chalcids play a relatively small role in gall formation in the West. They are much more

Plate 17.
A chalcid gall on catclaw caused by *Tanaostigmodes* sp.

prominent, however, as parasites of gall inducers. Be that as it may, two chalcid galls are described in this guide occuring on cat-claw *(Acacia greggii)* in the deserts of the Southwest.

References: Mani 1964; Felt 1965; Powell and Hogue 1979; McGavin 2002; Cranshaw 2004.

Sawflies (Order Hymenoptera, Family Tenthredinidae)

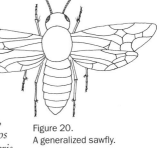

Sawflies (fig. 20) are actually diminutive wasps (less than 8 mm long) that get their strange name from their sawlike ovipositors. Three genera are considered here on willows, the leaf-galling *Pontania* spp. and the stem-galling *Euura* spp. (pl. 18), and on snowberry (*Symphoricarpos* spp.), the bud-galling *Blennogeneris*

Figure 20.
A generalized sawfly.

sp. A fourth group, which includes *Phyllocolpa* spp., develops leaf-edge roll galls in willows, but are not covered here. Not all sawflies, however, are gall inducers. The two genera that do initi-

ate galls, on willows, for example, have distinctly different habits, which also contrast with those of the cynipid wasps. With sawflies, gall formation begins with the secretions released by females as they oviposit their eggs into plant tissue. In some cases (*Euura* spp.), galls are well developed before the eggs hatch, yet the larvae contribute to further gall development through

Plate 18. Integral stem galls on willow caused by *Euura lasiolepis*.

their feeding and chemical actions. Adult sawflies have biting-chewing mouthparts, which are particularly important to some members of the *Euura lasiolepis* group who chew their way out of their woody stem galls following pupation inside the galls. In contrast, the last-instar larvae of the *E. exiguae* group cut exit holes and plug them with frass (feces) before retreating back down into the gall for pupation. Adults push the frass out to emerge. Those members of the *E. lasiolepis* group normally associated specifically with willows subject to flooding in streambeds usually do not cut exit holes before pupation, which might allow them to survive flooding. Following emergence, *Euura* males and females will drink water and feed on willow stamens, pollen, and nectar. *Euura* sawflies produce a single generation each year. These sawflies are not only host specific but also show a preference for specific clones within a species of willow.

Some reports indicate parthenogenetic reproduction among certain species. The rather small adult females are 4 to 5 mm long and are responsible for gall formation since the galls are fully developed before the eggs hatch. With some species, larvae contribute to gall growth, but the adults are the primary gall inducers.

Most members of the genus *Pontania* also produce a single generation each year. In some locations, however, the Willow Apple Gall Sawfly *(P. californica)* breeds year-round, producing as many as six generations. These larvae press their voided feces into the walls of the larval chamber, maintaining a clean, frass-free feeding area. Ultimately, the pressed frass is incorporated into the gall tissue. Since the larvae have the ability to "mine" gall tissues right out to the thin epidermal layer, they often backfill those vulnerable tunnels to the epidermis with pressed frass. Other species apparently eject their frass from time to time through holes, later used by the prepupae to escape the galls. With the Willow Apple Gall Sawfly, the prepupae exit the gall, drop to the ground by silk threads, and spin cocoons within the leaf litter. These sawflies overwinter as prepupae. They pupate and emerge in spring in tune with the development of willow leaves. Adults live for 35 to 45 days. Strangely enough, it has been reported that *P. viminalis* galls in Europe (these also occur here) develop adventitious roots from the fleshy interior walls of the gall larval chamber. These roots grow out of the gall through the exit holes after the galls have dropped to the ground with leaf fall.

What happens next is not clear. Both genera of sawflies, *Pontania* and *Euura*, are usually associated with willow trees.

One of the great surprises in my outdoor explorations was the discovery of a lone, near-prostrate willow, covered with bright red *Pontania* galls growing on the vast outwash plain of the Baird Glacier, just north of Petersburg, Alaska. This boulder- and gravel-strewn plain was barren except for scattered sapling Sitka alders *(Alnus sinuata)*, mosses, and some lichens and one willow. And, yet, here the willow and its guest sawflies were united in their own isolated, micro-world, and neither the worse for the chance encounter.

References: Caltagirone 1964; Mani 1964; Smith 1968, 1970; Meyer 1987; Weis et al. 1988; Shorthouse and Rohfritsch 1992; Nyman 2000.

Cynipid Wasps (Order Hymenoptera, Family Cynipidae)

The galls of cynipid wasps (fig. 21) were among the first recognized for their commercial and medicinal uses by early naturalists as far back as 460 B.C. While uncommon in Asia and some other parts of the world, cynipid wasps are the dominant gall inducers in North America, with over 800 species known, most of which occur on oaks (pl. 19). Several others induce galls on members of the rose family such as wild roses (*Rosa* spp.) and thimbleberry *(Rubus parviflorus)*. Cynipids are also found on chinquapin (*Chrysolepis* spp.), and one species occurs on the weed cat's-ear (*Hypochaeris* sp.). The majority of galls arise on leaves and buds. Fewer galls occur on stems, branches, flowers, acorns, and roots. Cynipid wasps are responsible for the most extreme galls in color and shape. Galls that look like miniature stars, sea urchins, golf balls, cups, saucers, clubs, teardrops, goblets, and bow ties are among the fascinating shapes that stir the imagination.

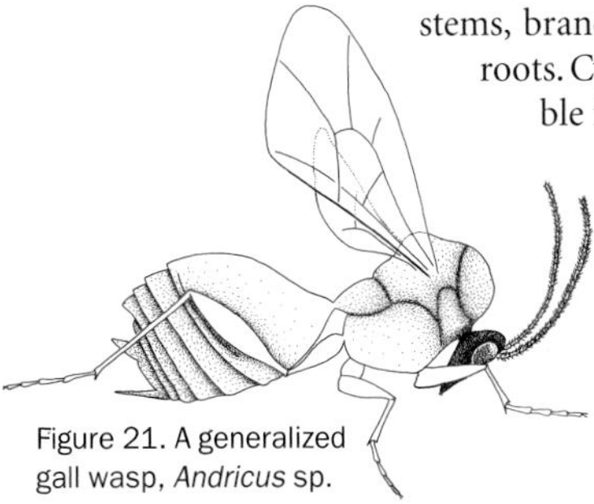

Figure 21. A generalized gall wasp, *Andricus* sp.

Plate 19.
The spangle gall of
Antron douglasii,
a cynipid wasp on
valley oak.

The shape, size, location, and color of cynipid galls are species specific, which often allows identification of the gall wasp without ever seeing the insect.

In cynipid galls, tannins and starch are absent from the nutritive tissues even though they occur in the outer layers of the galls, which may (in the case of tannins) serve as a defense against some penetrating enemies. Tannins may also suppress invasion by pathogenic fungi that could kill wasp larvae. Cynipid wasp galls are either monothalamous or polythalamous. Dozens of gall insects, inquilines, parasites, and hyperparasites can emerge from a single, large, polythalamous California Gall Wasp *(Andricus quercuscalifornicus)* gall over a three- to four-year period. While most cynipid wasps are gall inducers, some act as inquilines, feeding on the gall tissues of another cynipid wasp, and sometimes creating their own chambers within the host galls.

In some species, for example, the California Gall Wasp, there are no known males in the population. The females reproduce parthenogenetically, producing exact clones of the female parent. Other species have males and females and reproduce sexually.

Still, several exhibit an **ALTERNATION OF GENERATIONS**, called **HET-EROGENY**, with a spring sexual generation emerging from a different gall (often on a different plant organ) than the summer-fall generation females (agamic generation). The latter females overwinter in diapause as prepupae (usually in the galls) and pupate and emerge in spring in timed harmony with the development of the preferred plant organs. The eggs produced and deposited by these females result in the larvae and galls of the spring sexual generation. This alternation of generations is rather rare in the animal kingdom, but seen in aphids and rotifers as well. For many years, scientists identified some cynipid wasps as separate species, without knowing they were simply alternating generations of the same species. Research by several scientists over the last 50 years has clarified many of these relationships, including the discovery of new ones.

Once eggs hatch, larvae commence feeding by rupturing cell walls and jarring them enough to cause the release of cellular juice, which is consumed by the larvae. The thin layer of nutritive cells forming the larval chamber walls is constantly renewed through larval stimulation. The larval chamber wall of cynipid galls is usually well defined, smooth, firm, and either round or oval, in contrast to the irregular larval chambers of cecidomyiids and tephritids. In contrast to the tenthredinid sawflies, cynipid larvae are the principal gall inducers, not the adults. Cynipid larvae have strong chewing mouthparts primarily used to break open cell walls to suck the juice or cell sap from nutritive tissues. Cynipid larvae release auxinlike substances that are believed to be responsible for gall growth very soon after hatching and continue to do so throughout their larval development. They are also able to convert plant starch into soluble sugar through enzymatic action. Cynipid larvae have intestines that are closed for most of their larval life; just before pupation the gut opens and emits liquid wastes, which may be absorbed by gall tissues, since there is no evidence of these products during the prepupae or adult stages. Cynipid larvae do not produce solid wastes that would otherwise foul the chambers. Pupation occurs mostly inside the galls. Adults gnaw their way out to just under the epidermal skin and sometimes wait there until climatic conditions are favorable before pushing through the skin to emerge. Cynipid adults are generally small, around 2 to 3 mm, live only about a week, and do not feed.

Cynipid galls are distinguished by the presence of solid, concentric zones of differentiated cells around the larval cavities. The outermost layer or zone is typical of epidermal tissue. Beneath the epidermis is a zone of parenchyma cells covering the so-called protective zone of sclerenchyma cells. It is within this sclerenchyma zone that the nutritive zone of small cells, rich in amino acids and other useful nutrients, form to create the larval chamber wall. This last feature, the specialized nutritive cells in the larval chamber, is unique to cynipid wasps, since other gall-inducing insects do not share this trait. See the related discussion in the section on cecidomyiids.

Among cynipids, there are, however, exceptions to this concentric zones concept. Several paper-thin cynipid galls exist in which the larval chambers are suspended in the middle of the galls, held in midair and supported by radiating fibers that connect to the epidermis of the gall *(Trichoteras vaccinifoliae, T. coquilleti,* and *Besbicus mirabilis)*. The larvae are sustained through nutrients transmitted from the epidermis through the radiating fibers into the nutritive cells of the chamber walls. Certainly, this configuration must nearly eliminate inquilinous insects since there is no gall flesh to eat, except for the thin skin. But, several questions arise over the survival value or benefit to the gall organisms for such an unusual arrangement. Since galls with this design seem to occur either at high elevations (more than 1,500 m [5,000 ft]) or in areas of extreme summer heat, does the air mass insulate the larvae and pupae from extremes of heat or cold? Does the suspension of the larval chamber make it more difficult for parasites? Someday, we may fully understand the complexities of such unique features.

References: Kinsey 1922; Weld 1952a,b 1957, 1960; Lyon 1959, 1963, 1964, 1969, 1970, 1993, 1996; Doutt 1960; Mani 1964; Brown and Eads 1965; Felt 1965; Rosenthal 1968; Dailey 1969, 1977; Rosenthal and Koehler 1971a; Dailey and Sprenger 1973a,b; Russo 1979, 1981, 1983, 1990, 1991; Dailey and Menke 1980; Meyer 1987; Shorthouse and Rohfritsch 1992.

THE DETAILED DESCRIPTIONS of galls and the biology of their inducers are arranged according to host plant. For example, all the galls occurring on alders are grouped together in one section, with the gall inducers listed in phylogenetic order. The host plants are arranged alphabetically by common name. This guide proceeds from trees to shrubs, with a final section on miscellaneous herbs, ferns, and ornamental plants.

TREE GALLS

Alder Galls

At least four species of alders occur in the Pacific states: Sitka alder *(Alnus sinuata)*, red alder *(A. rubra)*, white alder *(A. rhombifolia)*, and mountain or thinleaf alder *(A. tenuifolia)*. Both Sitka and mountain alder can appear as small trees or shrubs, whereas mature red and white alders tend to be relatively tall. For the most part, none of the six gall organisms described here seem to be restricted to any one species. The gall organisms of alder include a nitrogen-fixing bacterium, a sac fungus, two mites, and two gall midges.

References: Mix 1949; Keifer 1952; Keifer et al. 1982; Sinclair et al. 1987; Gagné 1989; Johnson and Lyon 1991.

ALDER ROOT GALL BACTERIA   *Frankia* spp.
Pl. 20

These bacteria induce root galls that are rarely noticed unless you make an effort to look for exposed roots near stream banks or pull a small sapling out of the ground. Nitrogen-fixing bacteria create galls and in them extract nitrogen from the atmosphere and fix this nutrient in a form that can be used by the infected plants. The 5 to 30 mm wide, round, tubercled, or knobby galls occur just under the surface and completely surround the trunk or are fixed along the roots of seedlings and saplings. Dozens of these galls can occur on a small seedling no more than 15 cm (6 in.) tall. Technically, these perennial nodules are called actinorhizae. The nodules may live on alders for three to eight years. The bacteria fix much more nitrogen than can be used in their own growth and maintenance. So, the remainder is translocated to other parts of the host plant. After affected roots die, nitrogen and other nutrients are recycled at the site and become available

Plate 20. Nitrogen nodules induced by *Frankia* spp. on an alder seedling found on the relatively barren outwash plain of the Baird Glacier in southeastern Alaska.

to other plants' feeder roots. Infection occurs through root hairs, which begin swelling at the point of entrance. A lateral root also develops at this site and ultimately turns into the nodule. This relationship is critical because it allows pioneer plants to get started and survive in nitrogen-poor soils, which can include avalanche and landslide sites, glacial outwash plains, and flood zones where nutrient-poor materials have been deposited. Strains of *Frankia* are host specific, occurring on antelope brush *(Purshia tridentata)*, bayberry *(Myrica pensylvanica)*, buffalo berry *(Shepherdia argentea)*, casuarinas *(Casuarina* spp.), ceanothus *(Ceanothus* spp.), mountain-mahogany *(Cercocarpus* spp.), Russian olive *(Elaeagnus angustifolia)*, wax myrtle *(M. californica)*, and other plants. See table 5.

ALDER CONE GALL FUNGUS *Taphrina occidentalis*
Pl. 21

This fungus induces broad, flat, twisting cone bract galls during late spring and early summer, depending on elevation. The extended cone bracts expand wildly into tongue-shaped enlargements 40 mm long by 5 mm wide. Infected cones rarely have only a few enlarged bracts. The infected cone can be obscured by dozens of these tongue-shaped galls. Sometimes an individual cone will have two or three eruptions of distorted bracts, while

Plate 21. Swollen female cone bracts of red alder caused by *Taphrina occidentalis.*

Plate 22. Bead galls caused by *Phytoptus laevis,* an eriophyid mite, on alder.

still showing the shape of the cone. The galls are usually green with red along the margins. Those fully exposed to sunlight may be all red. Old, dark brown, infected cones may remain on the trees for years. As cones develop near their full size, the serpentine enlargements of the infected bracts begin protruding. Cone galls on alders at low elevations develop by late spring, whereas those on alders at high elevations develop later. Usually by August, cone galls in the Sierra, for example, have matured and started to turn brown and wither. The spores of this sac fungus survive winter and reinfect the host trees in spring. This fungus resembles yeast during the asexual part of its life in winter. This fungus occurs throughout the western United States. It is common among native alders and those used for garden and neighborhood park landscaping.

ALDER BEAD GALL MITE   *Phytoptus laevis*
Pl. 22

This mite induces round, green, yellow, red, and reddish brown bead galls on both surfaces of the leaves. These small, 2 mm wide

Plate 23. Erineum galls caused by *Acallitus brevitarsus,* an eriophyid mite.

bead galls reveal their rough surface under close inspection with a hand lens. As with other bead galls induced by mites, there are corresponding hair-lined openings on the opposite side of the leaf. These galls appear singly or in large numbers per leaf. The galls do not develop until after the leaves have reached full size. The adult mites that emerge from these summer galls spend winter in the fissures of the host trees' bark. No significant damage is done to the infected leaves. Although this mite does go through an alternation of generations, little else is known of its biology. It was first described in Europe and is now widespread in the western states, including Alaska.

ALDER ERINEUM GALL MITE *Acalitus brevitarsus*
Pl. 23

This mite creates white to cream-colored erineum pockets on the underside of alder leaves. These 3 to 5 mm wide hair-covered convex galls are quite different from most erineum pockets, which tend to be smaller and brown. You will notice the white to cream hair pockets on the leaf undersides, or you will see the corresponding bumps on the upper surfaces of the leaves. They almost always occur in large numbers per leaf, distorting the affected leaves enough to visually separate them from noninfected leaves. I have collected the galls from red and Sitka alders (*Alnus rubra* and *A. sinuata),* but they may occur on other species throughout the range of these hosts.

Plate 24. The midrib fold gall of *Dasineura* sp. A on alder.

Plate 25. The swollen bracts of male catkins on alder induced by *Dasineura* sp. B.

ALDER FOLD GALL MIDGE   *Dasineura* **sp.** *A*

Pl. 24

This midge creates long midrib folds on leaves that protrude on the underside. The swollen galls consume one-half to three-fourths of the length of the leaves. Galled leaves appear pinched along the midrib vein when you look at the **DORSAL** surface. The gall becomes quite noticeable when seen from the underside because it protrudes down about 10 mm. The gall midge attacks young leaves in spring. The larvae spin cocoons within the galls and pupate there. These galls have been recorded on red alder (*Alnus rubra*) from British Columbia to California and are likely elsewhere.

CATKIN GALL MIDGE   *Dasineura* **sp.** *B*

Pl. 25

This midge induces globular, lumpy polythalamous galls on the male **CATKINS** of red alder (*Alnus rubra*) and probably other alders as well. The large, swollen, puffy bracts of the gall are much larger than normal catkins. Galls measure 15 mm in diameter by up to 25 mm long. Normal, unaffected catkin tissues usually hang below galled tissues. The surface of the galls appears smooth, with well-defined jointed bracts that fit together snugly. The interior larval chambers are branched and interconnected. Galls collected in September issued adult Catkin Gall Midges two weeks after collection. If females also emerge under natural conditions in September, then they may lay their eggs in next year's catkin buds. Other galls collected at the same time of year had been thoroughly parasitized by eulophid wasps (Eulophidae). Little else is known about the biology of this new species.

Ash Galls

There are two distinct groups of ash trees in the western states. One group, referred to simply as "ash," is in the genus *Fraxinus,* family Oleaceae, while the second group, though called "mountain ash," is in the genus *Sorbus,* family Rosaceae. Oregon ash *(F. latifolia),* is known to support an eriophyid mite, *Eriophyes fraxinivorus,* that induces galls on the flowers. While I have not found this in the field, I have found an eriophyid mite galling the leaves of a mountain ash *(S. scopulina),* as described here. The taxonomy of ash mites is complicated and requires much more research to clarify.

References: Keifer et al. 1982; Johnson and Lyon 1991.

MOUNTAIN ASH LEAF   *Tetraspinus*
GALL MITE   *pyramidicus*
Pl. 26

This mite produces small, yellow leaf blisters on the dorsal surfaces of leaflets with corresponding brownish, hair-lined depressions on the underside. The galls measure 1 to 5 mm across, with some individual galls coalescing to form larger clusters. The yellowish green bumps on the dorsal surface of the leaflets are slightly raised. Each leaflet can contain dozens of these erineum galls. Though the specimens photographed in the field were on *Sorbus scopulina,* it is highly likely that other species of mountain ash support this mite.

Both *S. scopulina* and *S. californica* are known to support a different eriophyid mite, *Phytoptus sorbi,* that induces larger, green blisters on the dorsal surfaces of leaflets. These galls project

Plate 26. The erineum galls of *Tetraspinus pyramidicus,* an eriophyid mite.

higher than those of the previous species. In California, *S. scopulina* is often heavily infested with this mite. A third eriophyid mite, *Phyllocoptes calisorbi,* causes large, white erineum galls on the undersides of leaflets of both species of mountain ash.

Cottonwood, Aspen, and Poplar Galls

Cottonwoods, aspens, and poplars (*Populus* spp.) are attacked by a variety of organisms including a fungus, three eriophyid mites (Eriophyidae), several aphids (Aphididae), a beetle (Cerambycidae), a leaf-mining fly (Agromyzidae), and a gall midge (Cecidomyiidae). Additionally, aspens and cottonwoods host a number of canker fungi that cause weeping, ruptured bark and swollen tissues, but these are not dealt with here. The tree species considered here include black cottonwood *(P. balsamifera* subsp. *trichocarpa),* Fremont cottonwood *(P. fremontii),* narrowleaf cottonwood *(P. angustifolia),* quaking aspen *(P. tremuloides),* and the introduced Lombardy poplar *(P. nigra-italica).*

Aphids are by far the most common gall inducers found on cottonwoods in the West. They usually occur on stems, petioles, leaf bases, or the leaves themselves. Their taxonomy and genetic relationships, however, are complicated and in need of clarification. Based on genetic analysis, there appear to be western "races" of eastern species, for example, that make species identification difficult and confusing. Some of the species names used here may change with continued research.

References: Mix 1949; Keifer 1952; Palmer 1952; Harper 1959, 1966; Lange 1965; Grigarick and Lange 1968; Alleyne and Morrison 1977; Keifer et al. 1982; Sinclair et al. 1987; Gagné 1989; Johnson and Lyon 1991; Shorthouse and Rohfritsch 1992; Blackman and Eastop 1994; Cranshaw 2004; Wool 2004.

POPLAR LEAF BLISTER GALL FUNGUS   *Taphrina populisalicis*
Pl. 27

This fungus causes erineumlike yellow depressions in between the veins on either surface of the leaves of black cottonwood *(Populus balsamifera* subsp. *trichocarpa)* and Fremont cottonwood *(P. fremontii)*. Convex, light green bumps correspond to the yellow depressions on the opposite side of the leaves. When the galls are developing, the depressions are pale beige in color. The yellow pigment develops as the galls mature and begin to sporulate. The pigment is actually in oil droplets in the asci, where the spores are produced. The depressions and corresponding bumps measure 5 to 15 mm long by 4 to 10 mm wide and are round to irregular in shape. Sometimes the depressions or bumps coalesce, creating a long chain running parallel to the main vein. The galls cause little damage to the leaves and host trees. These galls can be confused easily with those of eriophyid mites or certain rust fungi. This fungus actually is globally distributed, occurring on a variety of poplars, as well as red willow *(Salix laevigata)*. The galls have been collected from Alaska to California.

POPLAR ERINEUM GALL MITE   *Phyllocoptes didelphis*
Pl. 28

This mite induces yellow green or pink-flushed, veiny, erineum lumps on the either surface of the leaves of quaking aspen *(Populus tremuloides)* in spring and early summer. The corresponding hair-lined depressions on the opposite side of the leaves are yellow reddish when fresh and brown when old. Some galls have hairs that are hardened lobes and not the typical hairs found in

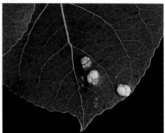

Plate 27 (left). Erineumlike pockets on the leaves of cottonwood induced by *Taphrina populisalicis*.

Plate 28 (right). Shallow erineum pockets on aspen leaves caused by *Phyllocoptes didelphis,* an eriophyid mite.

other erineum galls. The galls measure 5 to 15 mm in diameter and rise above the surface by 2 mm. The exaggerated veins of the leaves show on the convex surface of the galls. Frequently, leaf tissue near the galls blackens as a result of the redirection of nutrients into gall tissue. Infestations seem to be largely on the lower, shaded branches of trees. The galls occur singly or in large numbers per leaf. Mites live within the hair-lined depressions and remain motionless when disturbed. Adults overwinter around the buds. This gall mite is extremely common and abundant in some areas and absent in others. Another mite, *Aculus dormitor,* is an inquiline in the gall.

POPLAR BUD GALL MITE   *Aceria parapopuli*
Pl. 29

This mite induces nodular, rough, minutely **PUBESCENT** bud galls on quaking aspen *(Populus tremuloides)* but also occurs on black cottonwoods *(P. balsamifera* subsp. *trichocarpa)* and Fremont cottonwoods *(P. fremontii)* in California and other western states. The galls usually form on axillary and terminal buds and often involve large sections of the stems. The galls may be quite abundant on the affected tree. This mite also produces large, singular

Plate 29. The large knobby bud galls of *Aceria parapopuli*, an eriophyid mite, on aspen.

bud galls on the branches of black cottonwoods, which are usually round and bright red. The galls measure up to 40 mm long by 25 mm across and may last on host trees for several years. Fresh galls are pale green with pink to reddish tints, but ultimately they become dark brown in age. Heavy infestations can cause noticeable deformations of petioles and branches, with some stunting occurring. In places where the growing season is short, the activities of large numbers of this mite can seriously impact the growth habits of stems, leaves, and flowers. It is not uncommon to find heavily infested trees growing right next to trees completely free of infestation. Much of the life cycle of this mite is unknown.

POPLAR VAGABOND GALL APHID
Pl. 30

Mordwilkoja vagabunda

This aphid induces the largest aphid galls in the West. These **CONVOLUTED**, bladderlike leaf galls occur on quaking aspen *(Populus tremuloides)* and other cottonwoods. Fully developed masses usually exceed 30 mm wide and are composed of the crumpled leaves of terminal buds. In spring, eggs hatch and young nymphs move to developing buds, where they begin sucking plant juices. This feeding causes normal leaf tissues to transform into irregular masses of swollen leaf tissue. The feeding fundatrices are literally trapped within the galls during the early stages of development. The aphids appear able to continue stimulating gall growth during their feeding. As the galls mature and the season progresses, the ends of small projections on the galls dry and split open, allowing escape. The alternate host(s) are not known, but by late summer the winged generation returns to cottonwood hosts and

Plate 30. The convoluted leaf gall of *Mordwilkoja vagabunda* on aspen.

inhabits old galls, where it lays eggs. These aphids overwinter as eggs in old galls or bark crevices and have multiple generations each year. Old galls may last on the trees through the next season. This gall and its aphids occur throughout much of the United States and especially from the Rocky Mountains to the West Coast.

POPLAR PETIOLE GALL APHID   *Pemphigus*
Pl. 31   *populitransversus*

This aphid induces fleshy, orbicular-spherical, singular green galls with transverse slits on petioles at the base of leaves on Fremont cottonwood *(Populus fremontii)*. It has also been found on black cottonwoods *(P. balsamifera* subsp. *trichocarpa)*, narrowleaf cottonwoods *(P. angustifolia)*, and quaking aspens *(P. tremuloides)* in some areas. The prominent, angular or transverse slits often protrude slightly, taking on the appearance of lips. A tiny opening develops in the center of each slit. Galls measure up to 14 mm in diameter. The galls are smooth, green red (depending on sun exposure), and usually located on one side of the petioles. The rapid growth of the petioles often causes the petioles to bend to a 90-degree angle.

The swelling gall begins its development around the stem mother as she feeds, which provides the stimulus for gall development. Eventually, the gall completely encompasses the stem mother. This female lays numerous eggs, filling the gall with hungry offspring. The white, waxy secretions of the young often conceal them. As the galls become overcrowded, the young exit and crawl about on the leaves and stems, sometimes moving into other galls. One study has shown that these aphids abandon their

Plate 31. The petiole gall of *Pemphigus populitransversus* on Fremont cottonwood.

home gall when the heavily infested host tree begins to prematurely drop galled leaves (as a defense) or when predators show up. Often the aphids seek other galls and crawl in to finish their development. Aphids are usually present in the galls all summer long. In fall, as the leaves change color, wither, and drop to the ground, petiole galls remain green and succulent for up to two weeks. Once the galls begin to dry, the transverse slits open, allowing escape of the aphids.

Along coastal rivers, the galls of this aphid develop at varying times depending on the height of winter flooding. Those sections of the trees above the high-water mark produce galls earlier than the lower sections subject to winter flooding. Flooding may destroy the overwintering aphids from the previous season's galls. Usually, galls on lower sections of host trees do not develop until aphids have moved down from already formed galls. Cottonwoods in the warm interior valleys develop galls earlier than do trees of cooler coastal valleys or high mountain areas. I have found this aphid gall usually on Fremont cottonwood, even though it has been reported on quaking aspen and other cottonwoods. Green Lacewing (*Chrysopa carnea*) insect larvae and several birds have been reported as predators of this aphid. Of greater importance to these aphids, however, is the presence of a fungus that develops within galls, causing a high mortality among early nymph stages. The alternate "root hosts" of this aphid include a number of cruciferous plants such as turnips (*Brassica rapa*) and cabbage (*B. oleracea capitata*).

Plate 32. The petiole galls of *Pemphigus populicaulis* on black cottonwood.

POPLAR GOUTY PETIOLE GALL APHID   *Pemphigus*
Pl. 32   *populicaulis*

This aphid induces globular, succulent galls on the petioles at the base of leaves of several cottonwoods especially black cottonwood *(Populus balsamifera* subsp. *trichocarpa)*. It may also occur on quaking aspen *(P. tremuloides)*. Unlike the preceding gall, a series of small holes develop later in the season along the spiral slit, through which aphids squeeze out. This gouty-looking gall occurs singly or paired on each side of the petiole. A slight twist to the petiole often develops at the point of juncture with the leaf. Shapes vary slightly from round to ovular, and there are often small bumps across the surface. The galls measure 10 to 15 mm long by 8 to 10 mm across. Ants have been observed chewing through the walls of the galls. On several occasions in late summer in the Sierra at about 1,500 m (4,900 ft), I have found ants within empty galls with no aphids present. Did the aphids leave upon arrival of the ants? Did the ants consume the aphids and their sweet exudate? These and many such questions remain unanswered. It has been shown that water parsley *(Oenanthe sarmentosa)* is the alternate host for this aphid.

Plate 33. The stem gall of *Pemphigus populiramulorum* on Fremont cottonwood.

Plate 34. The fold or pouchlike galls of *Pemphigus populivenae* on black cottonwood.

STEM GALL APHID   *Pemphigus populiramulorum*
Pl. 33

This aphid induces integral stem galls on new growth of Fremont cottonwood *(Populus fremontii)* and black cottonwood *(P. balsamifera* subsp. *trichocarpa)*, usually just below a leaf-bearing petiole. The vertical slit, which distinguishes this species' galls, sometimes has slightly protruding lips. The slit measures about 7 mm long, while the entire gall measures 10 mm in diameter. A few specimens have reached 25 mm in diameter. The fundatrix leaves the gall between July and September for an unknown secondary host. This species has been recorded in several locations in western North America, notably Colorado, Utah, South Dakota, and California and may have several races or subspecies in these areas.

COTTONWOOD LEAF GALL APHID   *Pemphigus*
Pl. 34   *populivenae*

This aphid, also known as the Sugarbeet Root Aphid, induces elliptical, wrinkled fold galls on the leaves of cottonwoods including Fremont cottonwood *(Populus fremontii)*, narrowleaf cottonwood *(P. angustifolia)*, and black cottonwood *(P. balsamifera* subsp. *trichocarpa)*. Often, the galls are located in rows along the midrib vein or along the edges of the leaves. The aphids actually feed on the underside, causing the leaf tissue to fold around them, creating corresponding bumps on the dorsal surface of the leaves. The pale green, yellow red galls measure 10 mm long by 3 to 5 mm wide. Linear slits form on the underside of the galls, allowing escape of the aphids later. With numerous galls per leaf, the individual leaves can be severely distorted.

There is some evidence that some clones of cottonwoods may be resistant to attack, while other clones are more vulnerable. This species has been recorded from Colorado to California and uses sugarbeet *(Beta vulgaris)* roots, lamb's-quarters *(Chenopodium album)*, and pigweed *(Amaranthus* spp.) as alternate hosts.

COTTONWOOD FLASK GALL APHID
Fig. 22

Pemphigus bursarius

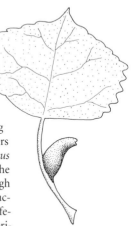

This aphid induces tubular, flask-shaped, midpetiole galls on Lombardy poplar *(Populus nigra-italica)* in California's Central Valley and elsewhere. One to four galls may be found on a single petiole. The galls measure 10 to 15 mm long. In the galls, the stem mothers can produce over 300 of their winged offspring, who later fly to secondary hosts, including lettuce *(Lactuca sativa)*, lamb's-quarters *(Chenopodium album)*, and carrot *(Daucus carota)*, and infest their roots. While on the secondary hosts, these aphids go through additional generations, ultimately producing the reproductive stage. The pregnant females of this generation return to the primary host Lombardy poplars to lay eggs that overwinter. This introduced aphid is widespread across North America and is native to Europe, where it is known as the European Lettuce Aphid.

Figure 22. The petiole gall of *Pemphigus bursarius*.

Plate 35.
The petiole gall of
Pemphigus sp. A.

PETIOLE GALL APHID   ***Pemphigus* sp. *A***
Pl. 35   ***(populiramulorum?)***

This aphid induces twisted petiole galls near the base of leaves of Fremont cottonwood *(Populus fremontii)*. Until recently, in most publications these galls were usually associated with *P. nortoni.* Recent DNA analysis of the adults and alates, however, has shown that the insects from these galls are closely related to the Stem Gall Aphid *(P. populiramulorum)* group, which may well have several western "races" or subspecies. The galls look superficially like those of the Poplar Gouty Petiole Gall Aphid *(P. populicaulis)* except that the exit hole with these galls is a spiral slit near the base of the leaf, and the galls are smooth. Galls measure 14 mm long by 10 mm across. Stem mothers can produce 700 to 1,000, or even more, offspring in each gall. Fungal infections that start in early June are the principle cause of in-gall mortality for this aphid. As with other aphid galls, lacewing larvae and select birds are predators of the aphids. Continued research may clarify the relationships of several undescribed relatives in the Stem Gall Aphid group in the West.

COTTONWOOD MIDRIB GALL APHID   ***Pemphigus* sp. *B***
Pl. 36

This aphid induces large, globular galls that protrude from the underside of the leaves on the midrib veins of Fremont cottonwood *(Populus fremontii)*. The dorsal surface of the galled leaves reveals a slight, discolored swelling highlighted by a linear slit

Plate 36. Left: An unusual midrib gall of *Pemphigus* sp. B associated with *P. populiramulorum* group, dorsal view. Right: The ventral view of the same aphid's gall showing prominent bulges.

that runs down the midrib vein for the length of the gall. The tissues around the slit are usually yellow in May. The globular, knobby, somewhat spiraled swellings forming the galls underneath measure 20 to 25 mm long by 12 mm across. Galls usually form near the base of the leaves extending out to the middle. The galls disrupt the flow of nutrients enough to cause yellowing and death of outer tissues of the leaves. In May, galls are full of winged offspring along with the fat stem mother. Recent DNA analysis has shown that aphids removed from these galls appear to be a western "race" of the Poplar Gouty Petiole Gall Aphid *(P. populicaulis)*, even though the galls of these two insects, as seen in pls. 32 and 36, are quite distinct. While there may be variability in the form of the galls as well as site selection, there may also be several undescribed relatives of the Poplar Gouty Petiole Gall Aphid in the West.

COTTONWOOD TWIGGALL BEETLE   *Saperda populnea*
Pl. 37

This beetle produces integral stem swellings on cottonwoods and willows (*Salix* spp.), particularly saplings. It is actually a round-headed woodborer. Galls of this beetle are recognized by the hemispherical cuts (scars) made by the adult beetles into which

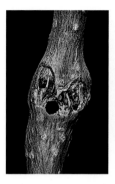

Plate 37.
Left: The gall of *Saperda populnea,* a longhorn woodboring beetle, with the exit hole of the adult.
Right: A newly emerged adult *S. populnea.*

an egg is inserted between the phloem and xylem tissues. Each gall measures from 12 to 20 mm across and is generally round to spherical, tapered gently, and not an abrupt swelling. Galls usually contain a single beetle larva (although two are occasionally found) that mines the interior wood, disrupting the flow of nutrients to outer areas of the branch. As a result, twigs weaken and are subject to breakage. Mature legless larvae measure about 25 mm long. On some trees nearly all branches greater than 15 mm in diameter are galled. Galls collected from Fremont cottonwood (*P. fremontii*) on the first of May issued adult beetles 14 days later. The adults are 10 mm long with antennae that are also 10 mm long. The dusty black adults, although capable of flying, tend to walk all over the branches of the host trees, constantly testing the air with their antennae.

POPLAR TWIGGALL FLY   *Hexomyza schineri*
Pl. 38

This leaf-mining fly induces large, abrupt, integral stem swellings on cottonwoods, poplars, and especially aspen. Saplings and young trees are particularly vulnerable to attack. Galls range from 20 mm to over 40 mm across and are characteristically abrupt swellings. These flies overwinter within the gall tissues as full-grown, yellow green larvae. Larvae develop into pupae within the galls, but the pupae work their way out of the galls and drop to the ground in winter or early spring. The 4 mm long adults emerge later, timed with the new growth of their host. During the day, flies rest and sun themselves. Impregnated fe-

Plate 38. The gall of *Hexomyza schineri*, a leaf-mining fly, on aspen.

males seek new growth and insert their eggs. Once the eggs hatch and the larvae begin feeding, the characteristic galls develop. Young branches can support dozens of these galls, sometimes in chains with one gall after another (see pl. 14). Individual galls usually contain two or three larvae, which become obvious by late summer and fall. This fly is particularly vulnerable to the parasitic chalcid wasp *Eurytoma contractura*, which normally accounts for 20 to 30 percent but can kill up 80 percent of the larvae in some years. Chickadees (*Poecile* spp.) are major predators of pupae. A *Cytospora* canker fungus often gains entrance to the gall through the exit wound of the fly. I have found a field of aspen saplings with nearly every sapling and every branch on the saplings supporting multiple galls. Once thought to be limited to the Rocky Mountains, these flies and galls are found in the Sierra Nevada of California and likely elsewhere.

ASPEN LEAF GALL MIDGE   *Prodiplosis morrisi*
Pl. 39

This midge induces rose pink to yellow green roll galls along the leaf margins of aspens. The galls often form along the entire edge of both sides of the leaves, creating a rippled, curled appearance. The roll forms inwardly toward the dorsal surface of the leaves. By mid- to late summer, these roll galls usually have a contrasting yellow green and adjoining dark brown discoloration. They are no more than 5 mm wide. Eggs are laid in newly developing leaf buds. Several larvae share the same space within the leaf rolls and drop to the ground to pupate. Galled leaves usually turn brown at the end of the larval cycle. Severe stunting can occur on infested terminal branches. This gall midge has been recorded on aspens

Plate 39. The leaf fold gall of *Prodiplosis morrisi* on aspen.

and some poplars from the East Coast to Kansas and Texas. Specimens shown here were collected from the central Sierra Nevada. This gall midge may occur elsewhere in the range of the host.

Cypress Galls

Two known stem galls are swellings induced by rust fungi on cypress (*Cupressus* spp.) (see table 6). Also three bud galls induced by gall midges occur on cypress in the West. One causes noticeable swelling of the branchlet tips, while the second induces small branch-tip galls. A third species induces large, swollen, four-sided bud galls.

Reference: Gagné 1989.

CYPRESS GALL MIDGE   *Walshomyia cupressi*
Pl. 40

This midge induces large, monothalamous, four-sided galls in summer on at least two cypress species in California. These galls form at the tips of branches of pygmy cypress (*Cupressus goveniana* subsp. *pygmaea),* within the stunted, bonsailike forests in Mendocino County, and the lofty Sargent cypress (*C. sargentii),* of Sonoma County. They appear as massive replicas of the terminal buds and often occur in clusters of three or four individual galls. The galls are 30 mm long by 10 mm wide. Each gall starts out as a swollen, beige, globular bud. But, as development proceeds, the overlapping scales separate from each other near

Plate 40. Three terminal bud galls of *Walshomyia cupressi*.

their tips. At this stage, the four-sided pattern is seen when you look down the axis of the galls. Galls at different stages of development can be found from spring through fall. Abandoned, old galls do not remain on the host plant for long. The adult gall midges emerge through the side near the tops after nearly two years in the larval stage within the galls. Adults live only for a few days in spring while they reproduce and lay eggs. The more vigorously growing cypress trees seem to host more galls than the more stunted specimens nearby.

TIP GALL MIDGE · *Contarinia* sp. A

Pl. 41, Fig. 23

This midge induces slight swelling of the branchlet tips of Nootka cypress *(Cupressus nootkatensis)* and pygmy cypress *(C. goveniana* subsp. *pygmaea)* and may occur on other species of cypress also. The monothalamous galls develop in spring and are light beige yellow when fresh. Galls measure 3 mm high and wide and are composed of swollen and distended leaf scales. The swollen bud scales flare away from the main stem, creating an open rosette or flowerlike pattern. These galls are often found in clustered groupings, with several adjoining branch tips affected.

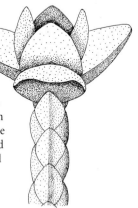

Figure 23. Comparative view of the gall of *Contarinia* sp. A.

Plate 41. The terminal bud galls of *Contarinia* sp. A on cypress.

Larvae pupate in the galls later in spring and emerge soon thereafter. There may be two generations per year, as fresh galls can be found in both summer and fall. This gall midge has been recorded in all Pacific states, as well as British Columbia.

PYGMY GALL MIDGE   *Contarinia* sp. *B*
Pl. 42, Fig. 24

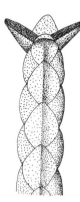

Figure 24. Comparative view of the gall of *Contarinia* sp. B.

This midge induces tiny, barely noticeable bud galls at the branch tips and occasionally along the sides of branches on Monterey cypress *(Cupressus macrocarpa)* and pygmy cypress *(C. goveniana* subsp. *pygmaea)*. Galls measure 1 mm high and wide and are the same diameter as the normal branches below, except the leaf scales forming the tip flare out, turn yellow or reddish brown, and reveal a central depression. These galls are detectable because normal terminal buds are rounded, while galled tips appear as tiny cups. Galls turn reddish brown after the larvae drop to the soil. Thousands of these galls can occur on a single small tree. This species is currently known only from California but is likely to be found in other Pacific Coast states where host plants occur.

Plate 42. The terminal bud gall of *Contarinia* sp. B on cypress.

Douglas-fir Galls

Douglas-fir *(Pseudotsuga menziesii)* is not a true fir. It is galled by a bacterium, a mistletoe, and at least three gall midges, one of which is described here in detail.

References: Furniss and Carolin 1977; Sinclair et al. 1987; Gagné 1989; Scharpf 1993.

DOUGLAS-FIR STEM GALL   *Undescribed*
Pl. 43

This is a globular, round, abruptly swollen gall that can exceed 30 cm (1 ft) across but are mostly under 10 cm (4 in.). Galls appear on twigs, branches, and trunks of young trees. The swellings cause eruption of the normal bark, which can open the tree to secondary invaders. Galls tend to appear in crowded stands on moist mountainsides. The gall inducer can suppress growth in young plants, resulting in dieback. These galls have been recorded from British Columbia to Arizona. Early research

Plate 43.
The knobby stem gall on Douglas-fir of unknown origin.

identified a bacterium, which was ultimately named *Agrobacterium pseudotsugae,* but later research has failed to confirm the identity of the causative agent. As Sinclair et al. (1987, 154) reported, "the disease is real, the pathogen is unknown."

DOUGLAS-FIR DWARF MISTLETOE   *Arceuthobium douglasii*

This mistletoe induces tapered swellings of branches and massive witches' brooms that easily exceed 1 m (3 ft) across even though the mistletoe itself is rather small and inconspicuous. The brooms are quite noticeable from a distance and, like other brooms, are composed of numerous tightly packed branches. Douglas-fir trees are severely damaged by this mistletoe. Heavily broomed trees are weakened, deformed, and often die. This is the only mistletoe found on Douglas-fir. These brooms have been found in Washington, Oregon, California, and Utah. On occasion, this mistletoe occurs on true firs (*Abies* spp.) and spruce (*Picea* spp.) growing in close association with infected Douglas-firs. See pl. 48 for a similar broom.

DOUGLAS-FIR NEEDLE MIDGE   *Contarinia pseudotsugae*
Pl. 44, Figs. 25, 26

This midge induces distinctive galls, usually at the base of needles. On the dorsal surfaces, galls appear as smooth, yellowish, succulent swellings, while on the underside the needle tissues swell around the larva with ridges of tissue growth. These galls

Plate 44.
The needles of Douglas-fir galled by *Contarinia pseudotsugae*.

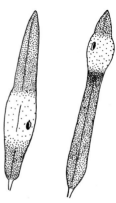

Figure 25 (left). Comparative view of the Douglas-fir needle galls of *Contarinia pseudotsugae*.

Figure 26 (right). Comparative view of the needle galls of *C. constricta*.

are only slightly swollen. Galls measure 5 to 10 mm long by 3 mm wide. While two other gall midges occur in Douglas-fir needles, this species is by far the most abundant. It has been recorded from California to British Columbia, Idaho, and Montana but may occur elsewhere in the full range of its host. This midge produces one generation annually, emerging from the soil in May to lay its eggs on the new needles. The yellowish larvae feed on the needles, inducing swelling to occur around the larvae. There may be several larvae per needle, with two or more galls occurring away from the usual location at the base of the needle. Full-grown larvae drop to the ground in late fall and early winter and then pupate in spring. Heavy infestations cause the needles to drop and twigs to die. This is a serious pest of trees grown on farms. The other two gall midges on needles of this host are *C. constricta* and *C. cuniculator*. Both occur from British Columbia to Montana but may occur elsewhere. *Contarinia constricta* has been found in California.

Fir Galls

A number of organisms gall true firs (*Abies* spp.), including rusts, mistletoes, adelgids (*Adelges* spp.), and gall midges (Cecidomyiidae). Refer to the lists of rusts (table 6) and mistletoes (table 8) for the species involved. Of the gall midges, most are associated with cone scales or seeds and are difficult, at best, to locate. I have chosen to focus only on a mistletoe, an adelgid, and a gall midge, which produce galls that are common and recognizable by the casual observer.

References: Gagné 1989; Johnson and Lyon 1991; Scharpf 1993.

WHITE FIR DWARF MISTLETOE
Pl. 45

Arceuthobium abietinum subsp. *concoloris*

This mistletoe produces tapered-elliptical swellings at the point of mistletoe growth on white fir *(Abies concolor)* and grand fir *(A. grandis)*. It occurs rarely on sugar pine *(Pinus lambertiana)* in the Sierra Nevada and on Brewer spruce *(Picea breweriana)* in southern Oregon. The galls caused by this mistletoe are typically 16 to 25 cm (6 to 10 in.) long. Older trees may show brooming as well as large cankers caused by a secondary invasion of *Cytospora* fungus. This mistletoe is common throughout the range of its hosts, from Washington through the Sierra to the mountains of southern California and along the coast into Mendocino County. A separate subspecies has been found on red fir *(A. magnifica)*.

BALSAM WOOLLY ADELGID
Pl. 46

Adelges piceae

This adelgid induces gouty stem swellings, usually near the tips of new growing shoots on fir trees. Sometimes, several galls occur on a single branch at previous years' terminal nodes. The galls are often seen on white fir *(Abies concolor)*, noble fir *(A. procera)*, grand fir *(A. grandis)*, and subalpine fir *(A. lasiocarpa)*, and the degree of damage varies greatly from species to species. Introduced from Europe or Asia around the turn of the twentieth century, it has spread throughout North America and is responsible for stunting and killing thousands of fir trees. It has been found in most western states. Its natural occurrence in California ap-

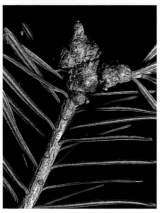

Plate 45 (above). The swollen stem gall of *Arceuthobium abietinum* subsp. *concoloris* on white fir.

Plate 46 (left). The terminal bud gall of *Adelges picea* on subalpine fir.

pears to be limited to red firs *(A. magnifica)* in northern California and other firs in Golden Gate Park, San Francisco, the East Bay Regional Parks Botanic Garden, Tilden Regional Park, Berkeley, and the University of California Berkeley campus. However, because this adelgid is a common pest of tree farms in the West, it is likely to occur elsewhere in the wilds of California. These aphid relatives are about 1 mm long and purplish to black. Only females are known in the United States. Eggs hatch into "crawlers" that are so small they are easily blown about by the wind. Usually, they crawl around until they find a suitable place to settle and begin feeding. Those blown away by the wind have a significantly reduced chance of locating a proper host and location for feeding. Once settled, the crawlers insert their long, sucking mouthparts into plant tissue and begin taking juice from the tree's tissues. White, wax ribbons are secreted from the sides and back of these feeding crawlers, ultimately covering them with a white woolly material. The swellings or galls develop from secretions inserted into plant tissues by these feeding adelgids and are often 20 mm or more in diameter and near the tips of branches. While minimal damage usually occurs to noble fir, heavy infestations of this galling organism can kill subalpine fir, even before the galls reach full size. You might first see these galls on fir trees sold in Christmas tree lots, although there is a growing trend to cut off infected tips prior to sale. Several predaceous syrphid fly larvae, as well as introduced lady beetles and aphid flies from Europe, are important biological controls of this adelgid.

FIR NEEDLE GALL MIDGE   *Paradiplosis tumifex*
Pl. 47

This midge induces swollen, elliptical, green beige galls near the base of needles of white fir *(Abies concolor),* balsam fir *(A. balsamea),* and Fraser's fir *(A. fraseri).* There can be more than one gall per needle. While most form near the base of the needles, some appear midway. Galls measure 6 mm long by 3 mm in diameter. They are usually green through at least midsummer but turn beige in fall. Females lay their eggs in spring when needles are growing. Larvae crawl to spots near the base of needles and begin to feed. With stimulation from the larva, the needle swells around each larva. There is only one larva per gall. Larvae usually leave the galls in the late summer or early fall and drop to the ground, where they enter diapause. This midge pupates the following spring. The galled needles usually die and drop off after the larvae leave. The Fir Needle Gall Midge has been found from the Canadian maritime provinces and Michigan south through the Appalachians to North Carolina and in California. It may also occur elsewhere along the Pacific Coast throughout the range of its hosts.

The life history of this gall midge has an interesting twist. The inquiline cecidomyiid *Dasineura balsamicola* lays its eggs shortly after the gall midge has laid its eggs and before the galls have formed. One or more of the larvae of *D. balsamicola* hang around the gall midge larvae until completely enclosed within the gall. Once within the gall, the inquiline larva feeds on the gall, growing

Plate 47. The needle galls of *Paradiplosis tumifex* on white fir.

faster than the larva of the gall midge, which eventually dies as a result of the inquiline. The larvae of the inquiline along with their "borrowed home" needles usually drop later than needles occupied just by the gall midge larvae. Pupation of the inquiline also takes place the following spring. This relationship has been reported from the Canadian maritime provinces to Michigan and California but may occur elsewhere throughout the range of its host midge.

Hemlock Galls

Two species of hemlock (*Tsuga* spp.) occur in the Pacific states: western hemlock *(T. heterophylla)* and mountain hemlock *(T. mertensiana)*. Each species of hemlock is attacked by a different species of dwarf mistletoe, resulting in witches' brooms. In both cases, the brooms are much more noticeable than the mistletoe.

Reference: Scharpf 1993.

WESTERN HEMLOCK DWARF MISTLETOE
Pl. 48

Arceuthobium tsugense subsp. *tsugense*

This mistletoe initiates massive witches' brooms that often exceed a meter across on western hemlock *(Tsuga heterophylla)*. Sometimes, entire forests of hemlock are infected with this mistletoe, leaving nearly every tree with multiple brooms. Also accompanying the brooms are spindle-shaped swellings and fasciations of the branches, but the brooms are the dominant feature. Broomed branches and trunks often split open, allowing invasion by secondary pests and fungi. Heavily broomed trees suffer limb loss,

Plate 48. A witches' broom gall caused by *Arceuthobium tsugense* subsp. *tsugense* on western hemlock.

stunting, deformation, and in young trees, death. While this mistletoe is rarely found on mountain hemlock *(T. mertensiana)*, it does occur on noble fir *(Abies procera)* and shore pine *(Pinus contorta* subsp. *contorta)* (on Orcas Island, Washington). I have seen an entire forest of western hemlock severely impacted by this mistletoe on the Brothers Islands in southeastern Alaska. This mistletoe occurs on western hemlock from Alaska to northern California.

MOUNTAIN HEMLOCK DWARF MISTLETOE
Arceuthobium tsugense subsp. *mertensianae*

This mistletoe induces witches' brooms on mountain hemlock *(Tsuga mertensiana)* that are similar to those of western hemlock dwarf mistletoe *(A. tsugense* subsp. *tsugense)*. Mountain hemlock dwarf mistletoe occurs in the higher elevations of the Cascade Range and the Sierra Nevada. At high elevations, where branching can be sparse due to strong winds, the numerous compact brooms stand out. Severe witches' broom formation and host mortality are characteristic of this mistletoe. Strangely enough, even though this mistletoe does not appear on western hemlock *(T. heterophylla)*, it does grow on noble fir *(Abies procera)*, subalpine fir *(A. lasiocarpa)*, and white bark pine *(Pinus albicaulis)*. It has been found from the central Sierra to central Oregon. See pl. 48 for a similar broom.

Incense-cedar Galls

Incense-cedar *(Calocedrus [Libocedrus] decurrens)* is found in the mountainous areas of California and Oregon. While Furniss and Carolin (1977) listed over 37 species of arthropods associated with incense-cedar, only two species are recognized here as gall-inducing agents: a rust fungus and a gall midge. A new, second species of gall midge has been discovered on our property in northern California during the production phase of this guide and is not described here. This bud gall midge has been tentatively placed in the genus *Walshomyia* until further classification studies are complete. A mistletoe, *Phoradendron juniperinum,* induces stem swellings on incense-cedar, but no brooms.

References: Furniss and Carolin 1977; Gagné 1989; Scharpf 1993.

INCENSE-CEDAR RUST   *Gymnosporangium libocedri*
Pl. 49

This is one of the most common rust fungi on conifers in the West. It alternates between incense-cedar and shrubs in the rose family, such as service-berry (*Amelanchier* spp.). This rust is occasionally found on apple (*Malus* spp.), pear (*Pyrus* spp.), quince *(Cydonia oblonga)*, and mountain ash (*Sorbus* spp.). On incense-cedar, it causes small to moderate-sized, erect witches' brooms generally less than 30 cm (1 ft) across. Even young seedlings of incense-cedar can develop brooms. The fungus may weaken the host tree or kill branches beyond the broom but rarely kills the tree, unlike some mistletoes. This fungus has an orange, gelatinous, **TELIAL STAGE** in spring during rainy periods. A heavily infected tree can be found practically dripping with blobs of orange, jellylike slime. Later, the sporidial stage is blown to a rose family host. There, cup-shaped fruiting bodies produce the aeciospores that infect new incense-cedar hosts.

INCENSE-CEDAR BUD MIDGE   *Walshomyia* sp.
Pl. 50, Fig. 27

This midge induces noticeable green, four-sided, terminal bud galls. The galls measure 10 to 15 mm long by 8 to 10 mm across, with flared, leafy bracts. Each bract is about 4 mm at the base and

Plate 49. A witches' broom gall caused by *Gymnosporangium libocedri* on incense-cedar.

Plate 50. The bud galls of *Walshomyia* sp. on incense-cedar.

Figure 27. The bract of the bud gall of *Walshomyia* sp. on incense-cedar.

9 mm long. A top-down view reveals the four-sided appearance. Larval chambers are located at the base of each leafy bract. While little is known about the biology of this gall midge, damage to the host appears negligible. At the present time, it is known only from California but may occur elsewhere in the range of its host. A second *Walshomyia* species gall midge has recently been discovered and induces heart-shaped bud galls.

Juniper Galls

Several species of shrub and tree junipers (*Juniperus* spp.) occur in the western states, including California juniper *(J. californica)*, western juniper *(J. occidentalis)*, Utah juniper *(J. osteosperma)*, and common juniper *(J. communis)*. With minor exceptions, there seem to be no species-specific restrictions among the several gall-inducing agents, which include rust fungi, mistletoe, and gall midges (Cecidomyiidae). The juniper mistletoe *(Phoradendron juniperinum)* is common on western juniper and causes noticeable stem swellings. Unfortunately, the gall midges involved with juniper are another group of gall insects whose taxonomy is confusing. The assignment of the genus *Walshomyia* is the best that can be done until further research clarifies the different species. Several species of rust fungi attack junipers and are listed in table 6. The midge galls are described here.

Reference: Russo 1979; Gagné 1989.

JUNIPER URN GALL MIDGE   *Walshomyia juniperina*
Pl. 51

This midge induces pea-sized, urn-shaped galls on the buds of California and western junipers *(Juniperus californica* and *J. occidentalis)*. The galls measure 7 mm wide by 10 mm high and have clasping lobes around the base. The three to four clasping lobes or bracts measure 2 to 3 mm wide and long at their bases. These galls often blend in with the normal cones of juniper. The galls generally have a silver gray bloom that develops with age. When fresh, these monothalamous galls are usually greenish gray to yellow brown, smooth, glossy, and closed at the apex. With age, the tips of the galls split open into three to five lobes, which recurve and flare out. This allows escape of the gall midges. Old galls are rusty brown, sometimes with a silver gray bloom, and wrinkled. Galls found in desert areas are fresh in March and April, while those on high mountain junipers are fresh in May and June. In July and August in the central Sierra, galls contain larvae in their last-instar stage. After this period, any open galls rarely have gall midge larvae, but instead the larvae of other insects. Hundreds of these galls can be found on a single juniper. As with other members of this genus, there is a single generation per year.

Plate 51. The bud gall of the *Walshomyia juniperina* on California juniper.

Plate 52. The bud galls of *Walshomyia* sp. A on Utah juniper.

JUNIPER GALL MIDGE — *Walshomyia* sp. A

Pl. 52

This midge induces conelike, monothalamous galls with open, reflexed or recurved bracts on the buds of Utah juniper *(Juniperus osteosperma)*. The apex of the galls is almost always colored with a silver gray bloom, while the lower bracts are olive green. Galls occur singly or in clusters of up to eight galls. Galls measure 14 mm high by 17 mm wide. An entire cluster may measure 30 to 35 mm across. The recurved bracts are 2 to 3 mm wide at their base. These midges have a single generation per year, with galls maturing in late fall and winter.

JUNIPER CONE GALL MIDGE — *Walshomyia* sp. B

Pl. 53

This midge induces robust, olive-sized, conelike galls on western juniper *(Juniperus occidentalis)* and has been found on several other species of juniper (Utah juniper *[J. osteosperma]* and California juniper *[J. californica]*). The galls, measuring up to 20 mm across, often occur in clusters of three to 18. Clustered galls are usually smaller. The galls are broad based but narrow toward the tips and composed of overlapping scales or bracts that are pressed together. These bracts do not have an open or recurved form but remain closed, except at the tip. When fresh, the galls are

Plate 53. The closed bud galls of *Walshomyia* sp. B on Utah juniper.

Plate 54. The bud gall of *Walshomyia* sp. C showing the flared opening at the apex through which the adult emerges.

olive green and often sticky but turn brown with age. Sometimes they have a silvery bloom. Little is known about the biology of this species.

JUNIPER TUBE GALL MIDGE   *Walshomyia* sp. C
Pl. 54

This midge induces elongated, tubular, monothalamous galls on the sides of the scaly leaves of Utah juniper (*Juniperus osteosperma*). Galls occur singly and are brown or light yellow with brown tips. These galls have been erroneously listed as deformed fruit. Galls measure 6 mm long by 3 mm wide. Occasional large specimens measure 8 mm high by 4 mm wide. Some have a slightly flared base, while others emerge right out of the bundled leaf scales. At maturity, the galls split open at the apex into three or four lobes with a central exit hole. Those galls that are parasitized do not split open at the apex. The parasite emerges through an exit hole on the side near the top of the gall. Little is known about the biology of this gall species.

Plate 55. The bud galls of *Oligotrophus juniperi* on Utah juniper.

JUNIPER BURR GALL MIDGE   *Oligotrophus juniperi*
Pl. 55

This midge induces rounded galls with flaring open bracts on the whorled leaves of Utah and California junipers *(Juniperus osteosperma* and *J. californica)*. Galls measure up to 20 mm wide and occur singly or in clusters. The bracts flare out from the sides of the galls, creating the burrlike appearance without recurving downward. The apex is open, with bracts spreading in all directions. Little is known about the biology of this species, as with most juniper gall midges.

Maple Galls

Eastern maples (*Acer* spp.) host a variety of gall-inducing erio-phyid mites (Eriophyidae) and gall midges (Cecidomyiidae). But, in the West the picture is different. While there are several species of maples in our area, only one species appears to support a gall organism. Mountain maple *(A. glabrum),* a delicate shrub-like tree fond of moist places, hosts a mite that produces a color-ful leaf gall. A white erineum gall caused by *Eriophyes negundi* on the undersides of the leaves of box elder *(A. negundo)* has been reported in the East, but there has been no report of it in the West.

References: Keifer et al. 1982; Gagné 1989.

Plate 56. The erineum galls of *Eriophyes calaceris* on mountain maple.

MAPLE ERINEUM MITE   *Eriophyes calaceris*
Pl. 56

This mite induces pink, magenta, and purplish red erineum galls on the leaves of mountain maple *(Acer glabrum)* from the Rocky Mountains to the West Coast and north into southeast Alaska. The erineum growth appears on the upper surfaces of leaves, mostly along the edges (although the erineum may cover entire leaves in some cases) and consists of brightly colored papillae. Each papilla is rounded at the tip and filled with a colored fluid. The rounded papillae can be seen without magnification. Other erineum galls usually have pointed hairs covering the feeding mites. The undersides of the leaves may be distorted, but not as severely as with other erineum galls. Sometimes, nearly all of the leaves of an individual tree may be covered with this noticeable erineum. This mite undergoes an alternation of generations. They continue to colonize the erineum galls until early fall, then migrate to bark crevices where they overwinter. In spring they move to developing buds. The eriophyid mite *(Aculops glabri)* is an inquiline found in the erineum.

Oak Galls

Twenty-four species of oaks *(Quercus* spp.), including hybrid crosses, are now recognized in California. Several varieties are also recognized. Oaks are classified into three basic groups:

BLACK OAKS: California black oak *(Q. kelloggii)*, coast live oak *(Q. agrifolia)*, and interior live oak *(Q. wislizenii)*

WHITE OAKS: valley oak *(Q. lobata)*, blue oak *(Q. douglasii)*, Engelmann oak *(Q. engelmannii)*, Oregon oak *(Q. garryana)*, Muller's oak *(Q. cornelius-mulleri)*, Tucker's oak *(Q. john-tuckeri)*, Nuttall's scrub oak *(Q. dumosa)*, MacDonald oak *(Q. macdonaldii)*, scrub oak *(Q. berberidifolia)*, leather oak *(Q. durata)*, shrub live oak *(Q. turbinella)*, and deer oak *(Q. sadleriana)*

INTERMEDIATE OAKS: canyon live oak or maul oak *(Q. chrysolepis)*, huckleberry oak *(Q. vaccinifolia)*, island oak *(Q. tomentella)*, and Palmer's oak *(Q. palmeri*, syn. *Q. dunnii)*

The Southwest has at least nine additional species, plus hybrids and varieties (see table 13, p. 132). With one exception (an eriophyid mite [Eriophyidae] found on both coast live oak and canyon live oak), cynipid gall wasps [Cynipidae] do not use oak hosts from more than one group. Gall wasps that use black oaks as hosts, for example, are not found on either white or intermediate oaks and vice versa. In most cases, entire genera of cynipid wasps are restricted to a specific group of oaks.

In the context of all the native species of trees and shrubs in the West, the oaks by far support more gall-inducing organisms than any other single group of plants. Lewis Weld (1957) listed over 110 species of cynipid gall wasps on Pacific Coast oaks and an additional 100 species on oaks in the Southwest (Weld 1960). It is noteworthy that nearly a dozen cynipids found on the canyon live oak in the Southwest also occur on the same oak along the Pacific Coast. Also, a cynipid on Arizona white oak *(Q. arizonica)* (and other related species in the Southwest) also occurs on scrub oak in northern California. Further research may reveal other duplications between these two distinct regional populations of oaks or with other species of *Quercus*, especially in southern California and northern Baja California. It also appears that a large number of little-known cynipids gall oaks in Mexico. For the purposes of this guide, I have excluded cynipid galls specific to Southwest oaks, simply because the diversity of these species and their host relationships are extensive and warrant a separate treatment.

Along with cynipid wasps, a fungus causes a witches' broom, a couple of eriophyid mites (Eriophyidae) produce erineum galls or brooms, and a few gall midges (Cecidomyiidae) produce leaf-fold galls on oaks, as described here.

The taxonomy of cynipids, as with other groups of gall inducers, is complicated, unclear in some areas, and controversial. A recent paper by Melika and Abrahamson (2002) has suggested a major consolidation of cynipid wasp species into larger genera, as well as the abandonment of some small genera. While I recognize this, albeit controversial, work here, I have chosen to use the established names and generic placements developed by previous cynipid taxonomists (Weld, Dailey, Dailey and Sprenger, Doutt, and Lyon, for example) until such time as name changes are clarified. In deference to Melika and Abrahamson's work, I have placed their proposed genus name changes in parentheses following the names currently in use in the species accounts, as well as listing the proposed changes that apply to this guide in table 12.

References: Weld 1952a,b, 1957, 1960; Lyon 1959, 1963, 1964, 1969, 1970, 1993, 1996; Doutt 1960; Dailey 1969, 1972, 1977; Dailey and Campbell 1973; Dailey and Sprenger 1973a,b, 1977, 1983; Dailey et al. 1974; Dailey and Menke 1980; Hickman 1993; Melika and Abrahamson 2002.

Black Oak Galls

The bulk of the galls described herein from the Pacific states are typically found on coast live oak *(Quercus agrifolia)* and interior live oak *(Q. wislizenii).* Fewer galls are found on California black oak *(Q. kelloggii)* in the field, even though this oak is listed as a potential host. The bulk of the gall inducers on the black oak group are cynipid wasps (Cynipidae). An unknown gall midge (Cecidomyiidae) creates leaf fold galls, while an eriophyid mite (Eriophyidae) develops erineum galls on leaves. The following galls are listed by the plant organ affected.

References: Weld 1957; Lyon 1959, 1964, 1969, 1970; Dailey 1969; Dailey et al. 1974; Russo 1979; Dailey and Menke 1980; Keifer et al. 1982; Sinclair et al. 1987; Melika and Abrahamson 2002.

TABLE 12 Proposed Changes to Cynipid Taxonomy
by Melika and Abrahamson

Current Name	Proposed Name
Andricus coortus	*Callirhytis coortus*
A. lasius	*Disholcaspis lasius*
A. reniformis	*D. reniformis*
A. spectabilis	*D. spectabilis*
Antron douglasii	*Cynips douglasii*
A. quercusechinus	*C. quercusechinus*
Besbicus conspicuus	*C. conspicuus*
B. heldae	*C. heldae*
B. maculosus	*C. maculosus*
B. mirabilis	*C. mirabilis*
B. multipunctatus	*C. multipunctatus*
Callirhytis congregata	*Andricus congregata*
C. carmelensis	*A. carmelensis*
C. perdens	*A. perdens*
C. perfoveata	*A. perfoveolata*
C. quercusagrifoliae	*A. quercusagrifoliae*
C. quercuspomiformis	*Amphibolips quercuspomiformis*
C. quercussuttoni	*Andricus quercussuttoni*
C. serricornis	*A. serricornis*
Dros atrimentus	*A. atrimentus*
D. pedicellatus (latum)	*A. pedicellatus*
Paracraspis guadaloupensis	*Acraspis guadaloupensis*
P. patelloides	*A. patelloides*
Sphaeroteras trimaculosa	*Atrusca trimaculosa*
Trichoteras coquilletti	*Andricus coquilletti*
T. rotundula	*A. rotundula*
T. tubifaciens	*A. tubifaciens*
T. vaccinifoliae	*A. vaccinifoliae*
Xanthoteras clavuloides	*Atrusca clavuloides*
X. teres	*Trigonapsis teres*

Note: This list includes some of Melika and Abrahamson's (2002) proposed
changes in cynipid taxonomy for species included in this guide. They have also
proposed changes for cynipids not in this guide.

Flower Galls

Three common galls occur on the flowers or catkins of coast live oak and interior live oak. To date, none have been listed for black oak.

SAUSAGE FLOWER GALL WASP
Fig. 28

Callirhytis congregata
**(*Andricus congregata*
Melika and Abrahamson)**

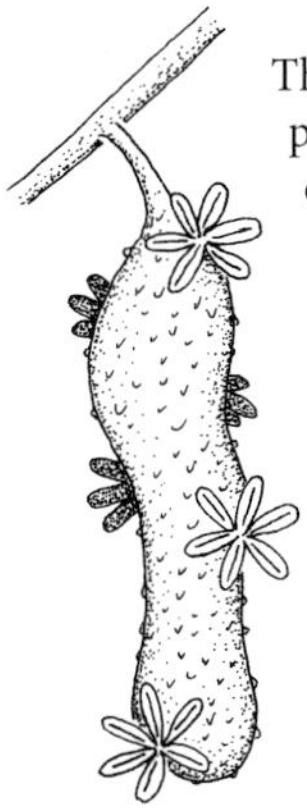

Figure 28. The catkin gall of *Callirhytis congregata* on coast live oak.

This cynipid wasp induces large, elongated, fleshy, polythalamous catkin galls, especially on coast live oaks. When fresh in April and May along the coast, the red and green, sausage-shaped galls stand out from normal catkins. Sometimes, these galls occur in clusters when all of the individual catkins in a group are galled. Generally, the entire catkin is involved in the development of the gall. Under these conditions, only the anthers protrude from the gall mass. The galls measure 40 mm long by 10 mm across. Young growing galls often release a sweet phloem exudate, which attracts ants and bees. At the peak of the population cycle for this wasp, as many as 90 percent of the catkins on an individual oak can be galled. Although adults have been reared from the galls in November, little else is known of the biology of this wasp.

KERNEL FLOWER GALL WASP
Figs. 29, 30

Callirhytis serricornis
**(*Andricus serricornis*
Melika and Abrahamson)**

This cynipid wasp induces clusters of hard, tiny, kernel-like, monothalamous galls of the bisexual generation on the flowers of coast live and interior live oaks. The individual galls normally occur in grapelike clusters at the upper end of the catkins. The entire mass may measure up to 20 mm in diameter. The individual, chocolate brown, glossy galls measure only 2 to 3 mm long. The obtusely pointed galls develop from the pollen sacs at the base of the flowers. Exit holes are conspicuous in spring, as males and females of this generation hatch out to reproduce and lay

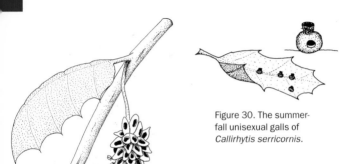

Figure 30. The summer-fall unisexual galls of *Callirhytis serricornis*.

Figure 29.
The spring bisexual galls of *Callirhytis serricornis* on coast live oak.

their eggs for the summer-fall generation. Females oviposit in the undersides of leaves in the veins. By July, the galls of this second, unisexual (agamic) generation are noticeable. These round, green and red galls are about 1 mm across, with a cap or collar on top of each rounded base. In the following spring, unisexual females oviposit in the unopened buds of staminate flowers, and the cycle begins again.

TWO-HORNED GALL WASP   *Dryocosmus dubiosus*
Pls. 57, 58

This cynipid wasp induces glossy, dark brown, club-shaped, monothalamous galls on the catkins and new leaves of coast live and interior live oaks. These bisexual generation galls are 2 mm wide by 4 to 5 mm long. Occasionally, two or more galls coalesce, creating what appears to be one large gall. Generally, however, they appear singly, or grouped together on a catkin or along the edges of new leaves. Galls on leaves generally cause indentations and, in some cases, can produce considerable malformation, making the leaves unrecognizable as leaves. Adults emerge in late April and live for about seven days. Females oviposit in the midrib and lateral veins on the underside of the leaves. After the eggs hatch, laterally compressed galls, looking like little pouches or purses, develop with a horn or point at each end. Fresh galls are green and measure 3 mm long. Dozens of these galls can occur on a single leaf. As they develop on the veins, they appear to reroute nutrients into the galls, depriving the leaf tissues beyond the galls of essential elements. This process results in the edges of the

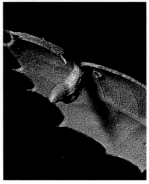

Plate 57. A single unisexual gall of *Dryocosmus dubiosus* on the underside of a coast live oak leaf.

Plate 58. Two bisexual generation galls of *D. dubiosus* on the edge of a coast live oak leaf.

leaves turning brown. When this wasp is at its peak population numbers, entire trees can appear as if they are burned or dying by September. The galls of this unisexual generation can be found on leaves well into fall and, often, through winter. The larvae overwinter in diapause and pupate the following spring when flower buds are developing. This cynipid wasp has an interesting competitive relationship with the California Oak Worm *(Phryganidia californica),* which also depends on oak leaves. The oak moth caterpillar eats the leaves that the cynipid wasp requires for egg deposition and gall formation. A full discussion of this relationship is in the section on environmetal factors. In spite of the tremendous ecological stress and competition between these two insects, they continue to survive using the same host trees.

Acorn Galls

While several acorn galls are known, only two species are described here. One occurs on coast live and interior live oaks, while the other occurs on California black oak. A third species, the Live Oak Petiole Gall Wasp (*Callirhytis flora* syn. *C. milleri*), uses acorns for its unisexual generation (see this wasp in the description of leaf galls on oaks in this group).

Plate 59. The honeydew secreting bud gall of *Callirhytis carmelensis* on coast live oak.

MOTTLED ACORN GALL WASP   *Callirhytis carmelensis*
Pl. 59   *(Andricus carmelensis*
                                Melika and Abrahamson)

This cynipid wasp induces glossy, mottled galls at the base of acorns of coast live and interior live oaks. I have not found them on California black oaks even though they are recorded as occurring on this species. Galls measure 6 mm high by 4 mm wide. They are maroon or green with light-colored mottling and small nipples at the apices. These detachable galls are monothalamous, laterally compressed, and not round and usually emerge from one side of the affected acorn, which is often severely stunted. The larval chamber is 2 mm across with a white nutritive layer. This gall produces honeydew that attracts ants, bees, and yellow jackets. See the section "Honeydew and Bees, Yellow Jackets, and Ants," in the introduction for more details.

ACORN CUP GALL WASP   *Undescribed*
Pl. 60, Fig. 31

This cynipid wasp induces galls that cause acorn cups on California black oaks (and perhaps other oaks in this group) to bulge out into distinct, rounded segments. These galled acorns stand out readily from nongalled acorns. The galls appear as individual segments of the cup and

Figure 31. A cross section of an acorn gall on black oak showing the larval chamber.

Plate 60. The gall of the Acorn Cup Gall Wasp on black oak.

usually measure 5 mm wide by 10 mm high. The larval chamber is 4 mm long and either centered in the gall or set toward the base. Oaks examined in mid-July on Mount Shasta in northern California were thoroughly affected by this gall insect.

Stem Galls

Four cynipid wasp stem galls occur on the black oak group. Two are detachable, and two are integral. Stem galls have the greatest capacity to disrupt the flow of essential nutrients, as discussed earlier. A powdery mildew fungus also attacks the buds of live oaks, creating massive witches' brooms.

LIVE OAK WITCHES' BROOM FUNGUS   *Cystotheca*
Pl. 61   *lanestris*
This fungus is a powdery mildew that initiates white witches' brooms on new shoots, which are commonly produced only in spring and early summer. The fungus is inactive in winter, so any new shoots produced then are not infected. This mildew fungus actually infects several species of oaks, including the coast live oak, canyon live oak, valley oak, and tanbark oak *(Lithocarpus densiflorus)* (the latter is not a true oak). This fungus is especially common in coastal areas. Infection starts with buds but usually results in a proliferation of shoots with dwarfed, whitish leaves and stems. These noticeable brooms can severely restrict spring growth in heavily infected trees. Infected trees often produce new shoots below the brooms, but these can also become infected. Brooms normally measure 20 cm (8 in.) or more across. In the south, spores survive winter on the surfaces of leaves while in the

Plate 61. The witches' broom caused by *Cystotheca lanestris,* a powdery mildew, on coast live oak.

north, spores reside amid buds. Unlike other fungi that require wet conditions, this fungus can do well even in low humidity and without condensed moisture and move systemically in young twigs. This mildew can penetrate leaf tissues with its hyphae. Many other mildews are superficial only, feeding on the epidermal cells of leaves and stems and sometimes flowers and fruit.

GOUTY STEM GALL WASP	*Callirhytis quercussuttoni*
Figs. 32, 33	*(Andricus quercussuttoni*
	Melika and Abrahamson)

This cynipid wasp induces round to globular, potato-shaped, woody stem galls. These integral galls are stem-colored, abrupt swellings that reach up to 10 cm (4 in.) long by 4 cm (1.6 in.) wide but are usually smaller. Older galls are often riddled with exit holes of the gall wasps, their inquilines, and parasites. These unisexual galls begin their development in summer and continue to grow until the following spring. Larval development terminates in late spring, with pupation taking place in the fall of the second year. Adults emerge from midwinter through March, nearly two years after the beginning of gall formation. Females of this unisexual generation oviposit in leaf buds. As the leaves unfold in spring, small, 3 mm long, green blisters develop on the petioles and leaf veins. From these monothalamous bisexual galls emerge

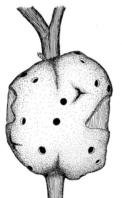

Figure 32 (above). The bisexual galls of *Callirhytis quercussuttoni.*

Figure 33 (left). The unisexual gall of *C. quercussuttoni* on coast live oak.

the males and females. Mated females normally oviposit in green twigs 3 to 5 mm in diameter from the previous year's growth. In summer, the integral stem galls of the unisexual generation begin development. Development of the unisexual galls usually causes the death of the branches beyond the galls. The extensive disruption of the flow of essential nutrients to the outer twigs and leaves caused by the galls is usually the fatal blow. In an oak woodland with many natural controls (predators and parasites), this gall wasp rarely causes any serious harm to the overall vigor and health of the host tree. Without natural controls in landscape situations, serious damage can occur.

RUPTURED TWIG GALL WASP *Callirhytis perdens*
Pl. 62 (*Andricus perdens*
 Melika and Abrahamson)

This cynipid wasp induces polythalamous, tapered, integral stem galls that rupture at maturity, expelling the kernel-like larval chambers. The galls measure up to 7 cm (2.8 in) long by 2.5 cm (1 in.) in diameter. While the galls are bark colored and gradually tapered, the linear black fissures that develop easily distinguish them from other stem galls. The fissures begin developing as the larvae mature. A dozen or more glossy, laterally flattened, beige larval capsules are forced to the surface and then drop to the ground. This exiting process usually takes place while the stems are still somewhat green and soft, usually after the first fall rains, and before the stems harden in fall. Once on the ground, the

Plate 62. The integral stem galls of *Callirhytis perdens* showing the openings for the larval chambers.

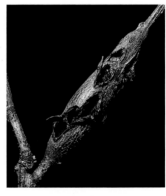

larval capsules nearly disappear amidst leaf matter and other debris. Even though the diapausing larvae in the chambers are somewhat protected by the shell of the capsules, birds and mice routinely eat the gall capsules and larvae. Sometimes, the larval capsules are carried off by ants, as in the case of the Two-horned Gall Wasp *(Dryocosmus dubiosus)*. An interesting advantage to this rupturing process, however it may be triggered or controlled, is that the adult gall wasp need only chew its way out through the thin capsule or chamber wall and not the thicker, hard tissue of the entire gall. On the other hand, the gall wasp becomes more vulnerable to predation when exposed to life on the ground. As with other integral stem galls, damage does occur to the branches beyond the galls.

LIVE OAK APPLE GALL WASP
Pls. 63–65

Callirhytis quercuspomiformis (*Amphibolips quercuspomiformis* Melika and Abrahamson)

This cynipid wasp induces two distinct galls belonging to separate generations of this species. Parthenogenetic females emerge from the previous summer's galls from January to March and oviposit in unopened leaf buds. As the leaves unfold, the alternate, bisexual generation galls begin developing on the underside of leaves. These galls appear singly or in groups of two or three. These mushroom-shaped galls measure 5 to 7 mm wide and high. When fresh, they resemble goblets or mushrooms with light green, yellow, red, and pink tones. With age, they turn beige. These monothalamous galls develop rapidly, usually within two weeks after oviposition. Males and females emerge from mid-May to early June. The females oviposit in stem buds, resulting in

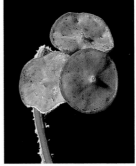

Plate 63 (upper left). The unisexual gall of *Callirhytis quercuspomiformis* on coast live oak.

Plate 64 (upper right). Three bisexual galls of *C. quercuspomiformis* showing the disc-shaped tops.

Plate 65 (left). The bisexual galls of *C. quercuspomiformis* showing their flared mushroom shape from the side.

the development of the summer, unisexual generation galls, which are large (to 3 cm across), round, and covered with numerous prickly spines or projections about 2 mm long. Specimens of these galls on coast live oak are often smoother than those on interior live oak. Fresh galls are succulent and green to red. With age, the galls fade to beige, and some sport black sooty molds on the surface. These polythalamous, pulpy galls have central larval chambers. The parthenogenetic females emerge the following spring. Robert Lyon, the researcher who discovered the unique relationship between alternating generations for this species, found evidence that suggested some of the parthenogenetic females laid eggs that produced only sexual females, while others laid eggs that produced only males. Lyon estimated that more than 75 percent of the bisexual, spring galls were parasitized.

LIVE OAK BUD GALL WASP
Callirhytis quercusagrifoliae
Pl. 66, Fig. 34
(*Andricus quercusagrifoliae*
Melika and Abrahamson)

Figure 34. The spring bisexual gall of *Callirhytis quercusagrifoliae*.

This cynipid wasp induces bright green, purple, beige, and white mottled, round, monothalamous bud galls. These minutely pubescent galls measure 6 to 9 mm in diameter. Larger specimens seem to occur on coast live oak compared to interior live oak. Development begins in early summer. In October and November, the now brown galls drop to the ground, sometimes by the hundreds. Many are taken by squirrels, mice, and wood rats. In February, parthenogenetic females emerge to oviposit in leaf buds. As the leaves unfold, tiny 2 mm long, one-celled blister galls develop on the leaves, in aborted buds, and on petioles. These galls develop rapidly, with the bisexual males and females emerging in April, seven to eight weeks after egg deposition. While the blister galls of the bisexual generation are difficult to see, the round galls of the unisexual (agamic) generation are easier to find.

Leaf Galls

Four common leaf galls occur on members of the black oak group in addition to the bisexual generation galls mentioned under the Live Oak Apple Gall Wasp *(Callirhytis quercuspomiformis)* and Live Oak Bud Gall Wasp *(C. quercusagrifoliae).* As a group, the black oaks in the West clearly do not support the diversity of leaf galls found on members of the white oak group or the intermediate oaks.

LIVE OAK ERINEUM MITE
Eriophyes mackiei
Pl. 67

This mite induces rusty brown, fuzzy, concave mats on the underside of the leaves of all live oaks in the black oak and intermediate oak groups. This crossover from one group of oaks to another is unique among oak gall inducers. The erineum pockets on the underside of leaves are normally rusty brown but can also be beige, pink, or rosy. The pockets on the underside have corre-

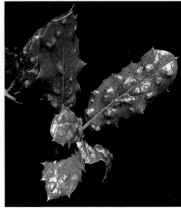

Plate 66. The terminal unisexual bud gall of *Callirhytis quercusagrifoliae*.

Plate 67. The erineum galls of *Eriophyes mackiei* on the dorsal surface of coast live oak leaves.

sponding convex bumps on the dorsal surfaces. Some leaves are so thoroughly infested with this mite that the entire undersurface of the leaves is covered with the expanded hairs, and the upper surface is considerably distorted. On the other hand, you can find small erineum pockets that are only 1 to 2 mm in diameter. The mites generally overwinter in the erineum pockets or among bud scales. New leaves are infested in spring and early summer. The expanded hairs of the erineum pockets are merely extensions of normal plant hairs. This mite is common throughout the West.

PUMPKIN GALL WASP   *Dryocosmus minusculus*
Pl. 68, Fig. 35

This cynipid wasp induces tiny, 1 mm wide, pumpkin-shaped galls on the upper surface of the leaves of coast live, interior live, and California black oaks. When fresh in late spring and summer, these monothalamous galls are light yellow with dark red or purple centers. As the galls mature and lose their color, they assume the appearance of Lilliputian-sized pumpkins

Figure 35. An enlarged version of the gall of *Dryocosmus minusculus*.

Plate 68. The gall of *Dryocosmus minusculus*.

with several lines or furrows radiating toward the depressed centers. In fall, most galls drop to the ground, where they blend in with leaf matter. Birds and mice eat some of these galls. Fresh and active galls can be found just about any time of year as you go from one region to another. Adults emerge the following April, and the cycle starts again.

LIVE OAK PETIOLE GALL WASP   *Callirhytis flora*
Figs. 36, 37   **(syn. *C. milleri*)**

This cynipid wasp induces large, hard, smooth-surfaced, irregularly shaped galls that usually engulf the entire petiole and sometimes part or all of the midrib vein of the leaves of live oaks in spring. These integral, polythalamous galls of the bisexual generation measure up to 4 cm long by 1 cm across.

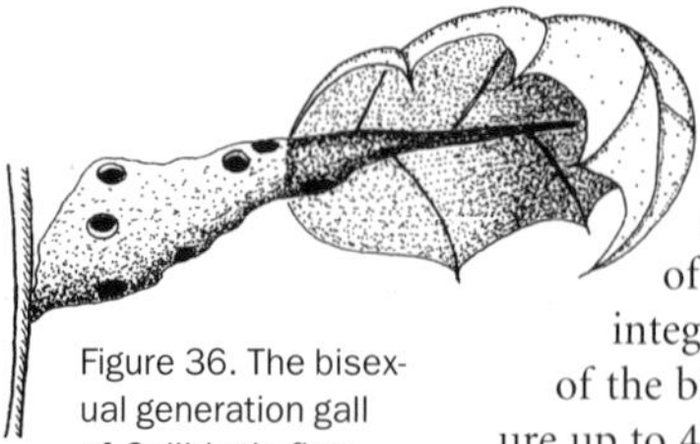

Figure 36. The bisexual generation gall of *Callirhytis flora*.

Galls show more on the underside of the petioles and leaves than on the upper side, which can be distorted by gall development. When fresh, the galls are green but soon turn glossy brown with age. From these galls males and females emerge in May and June. Impregnated females deposit their eggs in young acorns. These galls were formerly classified as belonging to a separate species *(C. milleri)*. In fall, the aborted, galled acorns drop to the ground, where the unisexual generation larvae remain in a pre-pupal stage for about a year and a half. Some of the females of this generation emerge from galled acorns to oviposit in petioles in February and March. Others delay emergence until the fourth or sixth year, with a few coinciding with the maturation of acorns. Once on the ground, squirrels and deer eat many of the acorns, while jays harvest other acorns for planting elsewhere as late winter food.

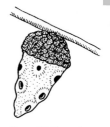

Figure 37. The unisexual generation gall of *C. flora.*

BALL GALL WASP
Fig. 38

Callirhytis perfoveata
(Andricus perfoveolata
Melika and Abrahamson)

This cynipid wasp induces round, monothalamous, integral leaf galls in spring on all species in the black oak group. The galls occur equally on both sides of the leaves and

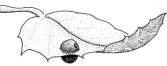

Figure 38. The integral leaf gall of *Callirhytis perfoveata.*

resemble balls surrounded by leaf tissue. When fresh, the galls are succulent and green; with age they turn brown and become hard. These smooth, glossy galls are usually about 10 mm in diameter. In cross section, there may be empty cavities next to the central larval chamber. Adult males and females emerge in April and May. Thus far, any alternate unisexual generation that might be expected to follow a spring cycle such as this has not been discovered.

White Oak Galls

White oaks compose the largest group of oaks in the western states, including a number of oaks in the Southwest (not covered here). As mentioned previously, the oaks in this complex group support the greatest number of cynipid wasps (Cynipidae) and the most extravagant galls in color and shape. The principle oaks from the Pacific states considered here are blue *(Quercus douglasii),* Muller's oak *(Q. cornelius-mulleri),* valley *(Q. lobata),* scrub *(Q. berberidifolia),* Nuttall's scrub *(Q. dumosa),* shrub live

(Q. turbinella), leather *(Q. durata),* and Oregon *(Q. garryana)* oaks. See table 13 for a partial breakdown of oak species. A full list for California can be found in *The Jepson Manual: Higher Plants of California* (Hickman 1993).

The majority of galls on oaks in this group occurs on the stems and leaves, including buds. Even though over 90 species of cynipid wasps have been described so far on stems and leaves, many more species and/or their alternate generations remain undescribed. While a few flower galls and a couple of acorn galls are known, they have been difficult to find. Therefore the bulk of the following descriptions concentrate on galls associated with buds, stems, and leaves.

References: Weld 1952a,b, 1957, 1960; Doutt 1960; Burdick 1967; Evans 1967, 1972; Rosenthal 1968; Dailey 1969, 1972, 1977; Rosenthal and Koehler 1971a,b; Dailey and Campbell 1973; Dailey and Sprenger 1973a,b, 1977, 1983; Burnett 1974; Dailey et al. 1974; Russo 1979; Dailey and Menke 1980; Gambino 1990; Heydon 1994; Melika and Abrahamson 2002; Schick and Dahlsten 2003.

Stem Galls

Stem galls on white oaks cannot compare to the variety and color of leaf galls, yet they are noteworthy nonetheless. Many bud galls are indistinguishable from other stem galls and are covered here in the same section.

IRREGULAR SPINDLE GALL WASP   *Andricus chrysolepidicola*
Pl. 69, Fig. 39

This cynipid wasp induces integral, polythalamous, spindle-shaped stem galls on blue, valley, scrub, and leather oaks. Galls of this unisexual generation measure up to 7 cm (2.8 in.) long by 2.5 cm (1 in.) in diameter. When fresh in the late spring and early summer, the galls are somewhat soft. By late summer and fall, they are sometimes knobby and extremely hard. These galls appear to cause little damage to branches and leaves beyond the galls, unlike other integral stem galls that completely divert nutrients into the galls. The abruptly swollen

Figure 39. The bisexual gall of *Andricus chrysolepidicola.*

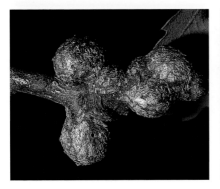

Plate 69. Several lumpy unisexual galls of *Andricus chrysolepidicola* on blue oak.

Plate 70. The terminal bud gall of *Andricus coortus* on Oregon oak.

galls are the same color as normal branches. Adult females chew their way out of the galls the following April or May and lay their eggs in buds. The bisexual generation galls, as described by Burdick (1967), develop rapidly. These ovoid, 2 mm long bud galls are covered with a fine **PUBESCENCE**. Males and females emerge, with females ovipositing into new stem growth. It is not uncommon for cynipids to have a short bisexual generation coupled with a rather long unisexual generation, as in this case.

CLUB GALL WASP *Andricus coortus (Callirhytis*
Pl. 70 *coortus Melika and Abrahamson)*

This cynipid wasp induces integral, polythalamous, round to globular, clublike galls on the tips of new stems of Oregon, blue, and scrub oaks. These galls form in spring and measure 13 mm high by 8 mm wide. They often have other normal and viable buds, sometimes stunted leaves, protruding from the sides of the galls. During the season after initial gall formation, buds that had protruded from the galls can elongate and become new stems. Two or three generations of galls may occur on the same branch. New galls are stem green with some red tones where they are exposed to direct sun. Adults emerge the following spring.

TABLE 13   Some Genetic Relationships between Common Oaks in the Pacific Coast and Southwest Regions

Pacific Coast black oaks	**Southwest black oaks**
California black oak *(Quercus kelloggii)*	Emory oak *(Q. emoryi)*
Coast live oak *(Q. agrifolia)*	Whiteleaf oak *(Q. hypoleuca)*
Interior live oak *(Q. wislizenii)*	

Pacific Coast white oaks	**Southwest white oaks**
Blue oak *(Q. douglasii)*	Arizona white oak *(Q. arizonica)*
Deer oak *(Q. sadleriana)*	Net leaf oak *(Q. reticulata)*
Engelmann oak *(Q. engelmannii)*	Gambel oak *(Q. gambelii)*
Leather oak *(Q. durata)*	Toumey oak *(Q. toumeyi)*
Nuttall's scrub oak *(Q. dumosa)*	Rocky Mountain blue oak *(Q. oblongifolia)*
Oregon oak *(Q. garryana)*	Shrub live oak *(Q. turbinella)*
Scrub oak *(Q. berberidifolia)*	
Valley oak *(Q. lobata)*	

Pacific Coast intermediate live oaks	**Southwest intermediate live oaks**
Canyon live oak *(Q. chrysolepis)*	Shrub oak *(Q. subturbinella)*
Huckleberry oak *(Q. vaccinifolia)*	
Island oak *(Q. tomentella)*	
Palmer's oak *(Q. palmeri)*	

CALIFORNIA GALL WASP   *Andricus quercuscalifornicus*
Pls. 71–74

This cynipid wasp induces the largest of the insect galls in the western states. These polythalamous galls can measure 12 cm (4.7 in.) long by 8 cm (3.1 in.) across. In spring, when fresh, these unisexual galls appear to burst out of supporting stems (although I have found fresh, emerging galls in September in some areas). They are succulent green or red, smooth, and glossy. By late summer, these lightweight oak apples turn creamy white and occasionally have a few short knobs or wartlike protruberances. By fall, the galls are beige. As the galls age they often support a black sooty mold and may stay on the host trees for three or more years. The galls occur singly or in clustered groups of a dozen or more. The general shape looks much like a potato, but under clustered conditions, the galls assume a pear shape.

Plate 71 (upper left). Early spring gall of *Andricus quercuscalifornicus.*

Plate 72 (upper right). The beige summer form of the galls of *A. quercus-californicus.*

Plate 73 (lower left). A cross section of the gall of *A. quercuscalifornicus* showing the adult female just prior to emergence.

Plate 74 (lower right). An adult female of *A. quercuscalifornicus* on the surface of a gall just after emergence.

This species is known only from parthenogenetic females. Neither males nor an alternate generation have been found for this species. The honey- or golden brown–colored females are among the largest of cynipid wasps, measuring 5 mm long. While some females emerge from August to November, others may not pupate and emerge for a year or more after gall formation, as influenced by environmental conditions; within the range of this species in the West, there seems to be some plasticity in the timing of emergence, oviposition, and new gall growth. Females deposit a dozen or more eggs per site in late fall directly into stem tissue, yet the galls do not begin development until the following spring. Generally, between March and May, stems rupture, revealing the rapidly growing galls. When first seen, the galls are about 15 mm long, but they reach full size usually within two months. Sometimes a second brood of galls begins development in July, even though the eggs were laid at the same time as those of the spring galls. The reason for the delay and the development of two separate broods is not clear. Strangely, the females from both broods appear to emerge at the same time even though one group has a head start. Larvae are clustered centrally in separate chambers, surrounded by a thick layer of pulpy tissue.

Old galls yield a variety of chalcid (Chalcidae), braconid (Braconidae), and ichneumon wasps (Ichneumonidae), as well as several other inquilines, parasites, and hyperparasites, over a three- to four-year period. Sapsuckers (*Sphyrapicus* spp.) may dig into these galls searching for larvae, leaving behind large holes. Southern California Indians made an eyewash from the galls. These wasps and their galls are common from southern Washington and Oregon through California into northern Mexico. This species occurs on nearly every species of white oak in the Pacific states, mentioned earlier.

ROUND HONEYDEW GALL WASP — *Disholcaspis canescens*
Pl. 2

This cynipid wasp induces densely pubescent, monothalamous, detachable, round, silvery, slightly mottled bud galls on blue and scrub oaks. The silvery gray bloom that covers the galls rubs off easily. The globular galls measure about 20 mm in diameter. Scattered tubercles are found on the surfaces of most galls. Females emerge in February and March. Fresh, growing galls are often found covered with ants, as these galls produce droplets of a sweet phloem exudate (honeydew) on the exterior (see the section "Honeydew and Bees, Yellow Jackets, and Ants," in the introduction). As mentioned earlier, several studies have shown a significant drop in parasitism rates along with a corresponding increase in gall wasp larval survival. Abandoned galls may remain on the host trees for some time after the insects have emerged.

Plate 75. The bud galls of *Disholcaspis conalis* on Oregon oak.

Plate 76. The bud gall of *Disholcaspis corallina* on blue oak.

Little else is known about the biology of this gall wasp or any potential alternate generation.

WITCHES' HAT GALL WASP   *Disholcaspis conalis*
Pl. 75

This cynipid wasp induces pointed, conical, **GLABROUS**, monothalamous stem galls on Oregon oak. These detachable galls develop in summer and occur singly or severally, separated along the length of a branch. Galls measure 15 mm high by 9 mm wide at the base. The sides of the galls are smooth. The base of the galls wraps over the stems in an undulating and flared manner. This gall belongs to a unisexual generation. Females are known to emerge in early fall. Neither an alternate generation nor males are known for this species. This wasp has been found in the central Sierra Nevada but is expected to occur elsewhere in the range of its host tree.

CORAL GALL WASP   *Disholcaspis corallina*
Pl. 76

This cynipid wasp induces round, monothalamous, orange, yellow, and reddish, detachable stem galls in summer on blue oaks. The key feature of these 10 mm wide galls is the presence of blunt,

Plate 77.
A single gall of *Disholcaspis eldoradensis* showing honeydew on the surface.

clublike, 2 mm long projections. The surface is minutely pubescent. These features create a resemblance to a South Seas coral head. With age the galls lose their colors, turn black with a sooty mold, and harden. They occur singly or in tight clusters of three or four at the base of leaf petioles. Adults emerge in fall. Even though this species induces one of the more attractive cynipid galls, little else is known of its biology.

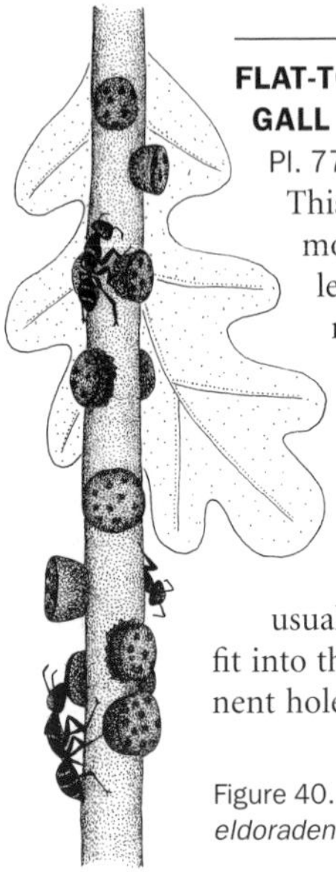

FLAT-TOPPED HONEYDEW GALL WASP
Disholcaspis eldoradensis

Pl. 77, Fig. 40

This cynipid wasp induces round to elliptical, monothalamous, detachable stem galls on valley, scrub, leather, and Oregon oaks in summer. These are the galls of the agamic or unisexual generation. The galls are generally elliptical with a flat or slightly convex top. The sides of the galls are glossy, sometimes wrinkled, and yellowish light brown. The tops of the galls are dull, nonglossy, pitted or fissured, rough, and dark brown. The galls usually have nippled attachments at the bases that fit into the stems. When pulled off the stems, prominent holes indicate where the nipples were attached.

Figure 40. An illustration of the galls of *Disholcaspis eldoradensis* with ants harvesting honeydew.

Plate 78.
The galls of
*Disholcaspis
mamillana.*

The galls usually reach their maximum size of about 8 mm by September. While the larvae are actively feeding and growing, the galls release a sweet phloem exudate that attracts large numbers of ants and especially yellow jackets (see the section "Honeydew and Bees, Yellow Jackets, and Ants," in the introduction). During late summer, these galls provide sugar to yellow jackets at a time when other sources are scarce. Galls may remain on the trees for several years, as evidenced by the presence of lichens on many specimens. Gall development occurs in rows, tight clusters, or singly. Adults emerge through holes in the sides of the galls in January and February. Evans (1972) described the bisexual generation as developing inside small, aborted, nondescript bud galls in spring.

BULLET GALL WASP   *Disholcaspis mamillana*
Pl. 78

This cynipid wasp induces round, monothalamous, finely pubescent bud galls on blue and scrub live oaks. The light beige galls may have a rose flush near the base and measure about 10 mm, with a short, 1 to 3 mm long, rounded nipple at the tip. The larval chamber is 3 mm long and connected to adjoining tissue. The larval chamber does not float freely in a larger cavity, as it does in the galls of the Dried Peach Gall Wasp *(D. simulata)*. The galls seem rather abundant in southern California, particularly in the San Luis Obispo area. Not much else is known about the biology of this insect.

TWIG GALL WASP
Disholcaspis mellifica
Pl. 79

This cynipid wasp induces elliptical, flat-topped galls that appear to burst out of cracks in the stems of scrub, leather, and Oregon oaks. The monothalamous galls usually appear in rows and often on several sides of the affected stems. Its sides are beige to red, with the tops pitted, rough, and darker brown or brick red. Galls measure 5 mm long by 2 mm wide and 3 mm high and are laterally flattened when viewed from the top. When developing in summer, these galls exude copious quantities of honeydew, which attracts ants, yellow jackets, and bees (see the section "Honeydew and Bees, Yellow Jackets, and Ants," in the introduction). One tree I found near Weed, California, supported thousands of these galls

Plate 79. Several galls of *Disholcaspis mellifica* showing drops of honeydew on the surface.

and the accompanying hordes of yellow jackets and ants on separate branches. This gall may be confused with that of the Flat-topped Honeydew Gall Wasp *(D. eldoradensis)* except for the swollen and ruptured bark on the stems and the consistent appearance in compact rows. Because the galls appear in summer and fall, with adults emerging the following spring, it suggests there may be an alternate generation. This could be, however, a parthenogenetic species, as with the California Gall Wasp *(Andricus quercuscalifornicus)*.

BEAKED TWIG GALL WASP *Disholcaspis plumbella*
Pls. 80, 81

This cynipid wasp induces one of the more spectacular galls in the West. The galls appear in two color forms: one is greenish overall with round yellow spots or lumps, and the other is deep wine red with bright yellow spots or lumps. These dramatic color

Plate 80. The red and yellow form of the galls of *Disholcaspis plumbella*.

Plate 81. The green and yellow form of the galls of *D. plumbella*.

patterns separate this species from all other look-alike bud galls, including those of the Bullet Gall Wasp *(D. mamillana)*. Additionally, the often long, curved, pointed beak at the opposite end of the point of attachment is a unique feature. A form of this gall appears in the Mojave Desert on shrub live oak, which has either a short, beaklike projection or none at all. The round, monothalamous galls of the Beaked Twig Gall Wasp occur on blue, scrub, scrub live, and leather oaks in late spring and early summer. The main body of the gall measures up to 15 mm in diameter. The otherwise solid gall has a central larval chamber. Gall development begins in May, in some areas, and reaches full growth by August. The galls remain on scrub oaks for several years, ultimately turning black with sooty mold. Adults are comparatively large, measuring 3 to 4 mm long. They emerge in November and December through exit holes in the sides of the galls. This gall wasp appears to be extremely common throughout the range of its host trees.

Plate 82. The galls of *Disholcaspis prehensa* with honeydew on the surface.

Plate 83. The phallic form of the galls of *D. prehensa*.

CLASPING TWIG GALL WASP   *Disholcaspis prehensa*
Pls. 82, 83

This cynipid wasp induces monothalamous, mushroom-shaped stem galls on scrub and leather oaks. The flaring sides of these galls are smooth and glossy, while the tops are rough, pitted, and nonglossy. The broadly clasping bases of the galls seem to wrap themselves around the stem, covering the points of attachment and concealing the stems. The sides rise to a narrow, constricted neck upon which sets the pitted cap. The galls usually measure about 10 mm wide at the base with caps about 5 mm wide. Altogether, the galls are 7 to 10 mm high. The sides of the galls facing the sun are often red, while the shaded sides or galls that are completely shaded are yellow, light green, or beige. These galls can occur singly or in tight rows and clusters on opposing sides of the stems. Another form of this gall occurs in the Mount Diablo area and is more phallic in shape with conical, pitted caps, little or no neck constriction, and the usual flaring, clasping base. These latter forms are often taller than the mushroom-shaped versions, ranging from 12 to 14 mm high. Both forms begin development in spring and reach maturity by early summer. Fresh, developing galls exude copious quantities of sweet phloem sap, which attracts hordes of ants (see the section "Honeydew and Bees, Yellow Jackets, and Ants," in the introduction). When removed from the stems, the galls leave deep pits where they were

Plate 84. The galls of *Disholcaspis simulata* on scrub oak.

attached. The elliptical larval chambers are located at the base of the otherwise solid galls. Adults emerge through holes in the sides in late winter and early spring. Smaller inquilinous insects pupate in and emerge from the caps at another time.

DRIED PEACH GALL WASP   *Disholcaspis simulata*
Pl. 84

This cynipid wasp induces monothalamous, round, densely pubescent stem galls on scrub, leather, blue, Oregon, and Engelmann oaks. These pointed or nippled galls look like small peaches. Galls often grow next to each other or singly. Galls measure 20 mm long (with nipple) by 17 mm wide. Color varies, with an undertone of green blushed with brick red and darker mottling. The pubescence does not rub off easily. Upon completion of larval growth, the larval chamber appears to separate from the slightly larger cavity wall, allowing the chamber to float freely. This unique feature is rare among cynipid wasp galls. What significance this separation has for the pupa or the later emerging adult is not known. Larval cells measure 3 to 4 mm in diameter inside a 5 mm cavity. Adults appear to emerge in winter. The galls of this species can be confused with those of the Bullet Gall Wasp (*D. mamillana*), except for the latter's absence of a free-floating larval cell.

Plate 85. The galls of *Disholcaspis washingtonensis* on valley oak.

Plate 86. The galls of *Besbicus conspicuus* on valley oak.

FUZZY GALL WASP   *Disholcaspis washingtonensis*
Pl. 85

This cynipid wasp induces round, fuzzy, detachable, monothalamous stem galls on almost all white oaks in the Pacific states. The 8 mm wide galls have a mealy-granular texture, usually have small knobs or bumps over the surface, and often have a short stalk or neck at the base. Color ranges from beige to dirty gray. These galls occur singly or in groups. The larval chamber is centrally located in the otherwise solid gall. With age, the brown galls develop a shallow, wrinkled surface texture. Adults emerge in November. Old galls can remain on the trees for a year or so after emergence of the adults. Beetles feed on old gall tissue, often hollowing out the interior to the point where the skin is only a skeletal reminder of the once solid gall.

ROUND GALL WASP   *Besbicus conspicuus*
Pl. 86   (*Cynips conspicuus*
Melika and Abrahamson)

This cynipid wasp induces detachable, monothalamous, solid stem galls on blue oaks that are sometimes confused with the preceding species, the Fuzzy Gall Wasp (*Disholcaspis washingtonensis*). The outer surface of the 8 to 10 mm wide galls is usually cov-

Plate 87. The fresh bud galls of *Besbicus heldae* on Oregon oak.

ered with short, beige to rusty brown hairs. Longer, whitish hairs may be sparsely interspersed with the short hairs. The inner gall tissue is usually a rich chocolate brown. Galls occur singly or in clusters that can grow squeezed together, forcing the slides to flatten and taper toward the bases. Single specimens are usually perfectly round. These galls do not possess a neck or stalk, as with the preceding species. Galls are fully grown by August. With age, the galls fade to a gray brown, barklike color. Adults emerge in late winter and early spring.

THORN GALL WASP   *Besbicus heldae* (*Cynips*
Pl. 87   *heldae* **Melika and Abrahamson**)

This cynipid wasp induces hard, spiny, detachable, monothalamous bud galls on Oregon and valley oaks. The spines covering the round to globular-shaped galls are flattened laterally, platelike, and often bent over and twisted rather than erect. Sometimes they appear randomly scattered across the surface of the galls. When these galls are fresh in summer and early fall, they are rose pink with a mealy-granular surface and measure 6 to 16 mm in diameter. Galls turn brown by winter after the larvae have stopped feeding. Galls occur singly or in clusters. Coalesced clusters can exceed 20 mm across. They may remain on the tree for several months. These are the galls of a unisexual generation. Little is known about the biology of this species.

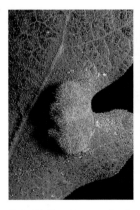

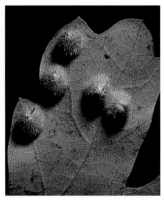

Plate 88. The erineum galls of *Eriophyes trichophila* on blue oak.

Plate 89. The urn-shaped galls of *Andricus albicomus* on the underside of an Oregon oak leaf.

Leaf Galls

Certainly the most spectacular galls on oaks in shape and color in the West occur on white oak leaves, particularly those in the Pacific states. These leaf galls reflect the ultimate variety in design and color within the gall-inducing insect world. Weld (1957) listed 47 species of cynipid wasp leaf galls occurring on white oaks in the Pacific states. In the intervening years a few additional species have been described, but certainly, many other new species are out there. Here, I include a few species that were originally pictured by Weld (1957) but have remained unidentified since (see the section "Unnamed Species" in the introduction). I also include new species not yet classified.

Some white oak leaf galls appear in spring, but others are summer-fall galls. In addition to cynipid wasps, an eriophyid (Eriophyidae) mite causes erineum galls on the leaves of blue oaks.

BLUE OAK ERINEUM MITE   *Eriophyes trichophila*
Pl. 88

This mite induces large, round to globular, erineum galls on the underside of blue oak leaves. The concave depressions are covered with whitish hairs, among which the mites feed. There are

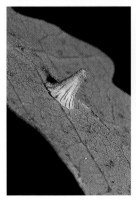

Plate 91. The spring bisexual galls of *A. atrimentus* on blue oak showing the transparency of the outer skin and revealing the inner, suspended, black larval chamber

Plate 90. The unisexual gall of *Andricus atrimentus* on blue oak.

corresponding convex bumps on the upper leaf surface. The bumps and corresponding depressions measure up to 15 mm across. These erineum galls are most noticeable during late summer and fall. Little else is known about the biology of this mite.

BRISTLE GALL WASP   *Andricus albicomus*
Pl. 89

This cynipid wasp induces monothalamous, round, bristly pubescent galls mostly on the underside and, sometimes, the dorsal surface of Oregon oak leaves. These white beige galls are covered with short, stellate hairs. These galls occur singly or in scattered groups along the edges of the leaves or near the midrib vein. Galls measure 4 mm high and wide. By October a small hole is usually apparent at the apex. Adults emerge in April, depositing their eggs in new leaves. This appears to be a northern California species, based on my current fieldwork.

STRIPED VOLCANO GALL WASP   *Andricus atrimentus*
Pls. 90–92

This cynipid wasp induces two unique galls on blue oak leaves: one a hollow ball, the other a striped cone. The bisexual genera-

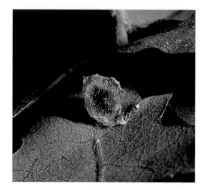

Plate 92. A cut away of the bisexual gall of *Andricus atrimentus* revealing the suspended black larval chamber.

tion galls are bulbous and hollow with a thin, pubescent, weblike skin. They have a dark blue black central larval chamber supported by dark fibers connected to the outer skin. When fresh in early March, these galls are light green to pink but ultimately turn beige with age. At this stage they are somewhat transparent, and the larval chamber can be seen through the thin outer tissue. Galls measure 3 to 4 mm in diameter. Females of this generation emerge within a two-week period in April and oviposit on the underside of leaves near the margin. Within 90 days after oviposition, the galls of the unisexual generation form. These volcano-shaped, cream-colored galls with red stripes are flat bottomed and flared at the base and measure 4 mm high and 4 mm wide. Parthenogenetic females emerge during late October through November and oviposit in leaf buds. After the new leaves open in spring, the galls of the bisexual generation develop and are well formed by March.

CLUSTERED GALL WASP   *Andricus brunneus*
Pl. 93

This cynipid wasp induces round, monothalamous, pubescent galls on the midrib on the underside of the leaves of most white oaks along the Pacific Coast. The galls usually occur in clusters, although single specimens can be found. They are dull lavender, pink, brick red to tan and are covered with short, cream-colored, matted hairs. The galls measure up to 6 mm in diameter. The outer wall of the gall is thick and hard. The larval chamber is cen-

Plate 93. A group of *Andricus brunneus* on the underside of a leaf.

Plate 94. A cluster of the galls of *Andricus confertus* on the underside of a valley oak leaf.

trally located and surrounded by lavender-colored cell tissue. The unisexual females emerge in fall. No alternate generation is known. These easily detached galls cause little or no damage to the leaves.

CONVOLUTED GALL WASP   *Andricus confertus*
Pl. 94

This cynipid wasp induces convoluted, brainlike midrib clusters of finely pubescent, detachable, monothalamous galls on the underside of the leaves of valley oak. The pink rose–colored gall mass is composed of individual, triangular galls measuring 3 to 5 mm in height and clustered on the midrib vein. The individual galls develop in such close proximity to each other, they look like one large mass measuring up to 14 mm in length. The lines creating the convoluted appearance actually separate each individual gall. There is usually only one gall mass per leaf. Gall development begins in early summer. Females generally emerge the following spring, but some do not emerge until a year or more after gall development.

Plate 95. A cluster of *Andricus crystallinus* galls on the underside of a blue oak leaf.

CRYSTALLINE GALL WASP   *Andricus crystallinus*
Pls. 95, 96, Fig. 41

This cynipid wasp induces caterpillarlike masses of bristly-haired galls on the underside of the leaves of blue, scrub, leather, and Oregon oaks in the late spring and early summer. Actually, brittle-haired gall masses are composed of small, individual, elliptical-spherical, monothalamous galls arising from separate points of attachment. Individual galls measure 12 to 14 mm high by nearly 7 mm across (including the hairs). Several growing in close proximity can together reach 35 mm long by 25 mm wide and completely encompass the entire underside of the affected leaf.

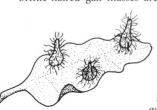

Figure 41. The bisexual generation gall of *Andricus crystallinus*.

Each individual gall usually has a slightly curved beak at the apex and a sparse coating of crystalline hairs. While the gall body may be a solid bright red, the hairs may be white, rose pink, red, or brown. The beaks often protrude through the hairy mass. These are the galls of the unisexual generation. In late winter, females emerge from these galls through exit holes near the tips of the beaks. They reproduce and lay eggs in leaf buds. About 17 to 20 days after oviposition, noticeable

Plate 96 (left). An individual
A. crystallinus gall.

Plate 97 (right). An old gall of
Andricus foliatus.

green, conical, slightly curved, monothalamous galls with cottony fibers develop on the upper surfaces of the leaves. These galls superficially resemble individual galls of the unisexual generation. The fibers are generally 2 to 3 mm and project laterally from the 3 mm high and 1 mm wide galls. These are the galls of the bisexual generation. Males and females emerge in March, with females ovipositing into the tissues of the underside of the leaves of the same hosts. Little damage seems to occur to the host's leaves even though summer-fall galls may cover leaves entirely.

LEAFY BUD GALL WASP   *Andricus foliatus*
Pl. 97

This cynipid wasp induces leafy, monothalamous bud galls on valley oak and, perhaps, other white oaks. Galls measure up to 25 mm long by 12 to 14 mm wide. In spring when these bud galls are fresh, they appear blasted with a compact collection of linear, thin, green, leafy bud scales or bracts. With age, the galls turn brown but are still easily ignored because they look like a normal plant organ. Little is known about the biology of this species, which may have an alternate generation.

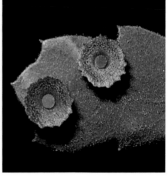

Plate 98. The fuzzy gall of *Andricus fulla-wayi* on the underside of a blue oak leaf.

Plate 99. The unisexual disc galls of *Andricus gigas* on blue oak.

YELLOW WIG GALL WASP　　　　　*Andricus fullawayi*
Pl. 98

This cynipid wasp induces kernel-sized galls covered with a dense mat of long, straw yellow– to fawn-colored hairs. These mono-thalamous, detachable, fluffy-haired galls measure 5 to 8 mm in diameter. The galls occur singly on the midrib vein on the under-side of the leaves of most white oaks. The long, soft hairs distin-guish this gall from all other hairy leaf galls on white oaks. The unisexual females emerge the following spring in March and April. It is not clear whether there is an alternate generation for this species. Little else is known about this species' biology.

SAUCER GALL WASP　　　　　*Andricus gigas*
Pl. 99, Fig. 42

This cynipid wasp induces small, 3 to 4 mm wide, slightly concave, disc-shaped, monothalamous galls on the upper leaf surface of blue, scrub, and Engelmann oaks in summer. The saucerlike unisexual generation galls are narrowly at-

Figure 42. The bisexual generation gall of *Andricus gigas*.

Plate 100. The bisexual generation gall of *Andricus kingi.*

tached at the base and have thin, flaring edges that are either smooth or toothed **(CRENATE)**. The larval chamber is a central bump in the otherwise concave gall. The surface of the gall is usually nonglossy. The color varies from shades of pink to red, purple, and brown. Under magnification, the galls have a short, matted pubescence. Galls occur singly or in large groups, with as many as 25 to 30 per leaf. Adult females emerge between December and February. By April, the bisexual generation galls develop on the staminate flowers and leaves. These galls are conical, tan, and about 3 mm high. Males and females emerge from these monothalamous galls in late spring to start the summer-fall unisexual generation. The unisexual generation galls may be confused with those of the Disc Gall Wasp *(A. parmula)* if not carefully examined.

RED CONE GALL WASP   *Andricus kingi*
Pls. 100–102

This cynipid wasp induces one of the more striking galls on white oaks. The red, monothalamous, detachable, cone-shaped galls occur on both sides of the leaves of blue, valley, and Oregon oaks in the summer months. The cone of these unisexual generation galls rises from a flared, cup-shaped base that is narrowly attached to the leaf. The apex of the gall, although pointed, is blunt tipped. The galls measure 5 mm high by 3 to 5 mm wide at the base. They occur singly or by the dozens on each leaf. I have found trees with thousands of these galls attached to nearly all of

Plate 101. The conical, unisexual generation galls of *Andricus kingi* with a gall of the Spined Turban Gall Wasp at the lower right.

Plate 102. A single unisexual generation gall of *A. kingi*.

the leaves. Under magnification, you can see a fine pubescence covering the entire surface of the galls. The larval chamber is large, occupying most of the base of the gall by early fall when the galls drop to the ground. Parthenogenetic females emerge through exit holes at the tips of the galls in February. Considerable controversy surrounded the bisexual generation galls, to the point of this species having been assigned incorrectly to *Liodora dumosae*. The bisexual, spring generation galls are actually small, monothalamous galls that form at the leaf margins. These 2 to 3 mm long galls taper to a rounded, narrow tip and turn black with age. Males and females emerge in May.

FIMBRIATE GALL WASP   *Andricus opertus*
Fig. 43

This cynipid wasp induces small, monothalamous galls on the underside of new spring leaves of scrub, leather, blue, and valley oaks. These are the galls of the bisexual generation that used to be referred to as *A. fimbrialis*, corrected by Evans (1972). These galls have a shredded appearance, with long, fiberlike projections emanating from the gall body. In some cases, the galls are extensions of the midrib veins. In other cases, the galls consume entire buds, completely disrupting the leaf. These galls are recognized by the long, stringlike or hairlike projections that emanate from the main gall body. Upon close examination, the veins of the host leaf can be seen as prominent ridges throughout the form of the

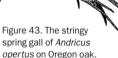

Figure 43. The stringy spring gall of *Andricus opertus* on Oregon oak.

gall. The gall can be 23 mm long, with the projections, by 6 mm in diameter at the widest point. The galls are leaf green when fresh but turn brown with age. Galls usually persist until fall, when they drop with their host leaves or on their own by wind action. Adults emerge in late spring. Like so many small, spring galls, those of this species can easily be overlooked as just a leaf deformity. The agamic or unisexual summer-fall generation develops in nondescript, aborted bud galls. Little else is known about this species.

DISC GALL WASP *Andricus parmula*
Pl. 103

This cynipid wasp induces monothalamous, flat, disc-shaped, glossy, detachable galls on both sides of the leaves of most white oaks. These galls measure up to 3 mm in diameter and have a narrow base of attachment and a smooth edge. The surface is often streaked with red and brown tones against a faint yellow. The upper surface of the otherwise flat gall is marked by a slight knob or **UMBO** over the centrally located larval chamber. These galls occur singly or in small clusters but are not confined to the main veins. Unisexual females emerge in April. Little is known about this species' biology.

SUNBURST GALL WASP *Andricus stellaris*
Pl. 104

This cynipid wasp induces convex, round, monothalamous galls with radiating crystalline projections on the underside of the leaves of blue and Oregon oaks. These unique galls measure 4 mm across by 2 mm high and are white, pink, and red in color. The central larval chamber is usually dark red. The crystalline

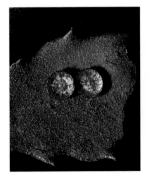

Plate 103. The summer gall of *Andricus parmula* on blue oak.

Plate 104. The galls of *Andricus stellaris* on the underside of a blue oak leaf.

projections are not hemispherical but instead radiate out laterally. When viewed from the side, the projections emanating from the larval chamber are separate and shorter than those that come from the base. The projections around the base of the galls sometimes appear club tipped. Galls occur in between the lateral veins, often along the margins of the leaves, either singly or in groups. Females emerge in spring even though an alternate generation has not been found. This species has been found in several counties within California but is expected throughout the range of its host trees.

STELLAR GALL WASP   *Andricus stellulus*
Pl. 105

This cynipid wasp induces stalked, cuplike, monothalamous galls on the leaves of Nuttall's scrub oak and shrub live oak. The most noticeable feature of the galls is the toothed cups surrounding the larval chambers, which resemble the galls of the Saucer Gall Wasp *(A. gigas)*, except for the prominent, hairlike stalks. These crenate cups, with their four to six toothy projections along the outer margin, sit on top of stalks connected to midrib or lateral veins on either side of the leaves, although I have found this gall attached only to the upper surface. There is a tiny collar around the

Plate 105. A fresh gall of *Andricus stellulus* on Muller's oak in the Mojave Desert.

stalk at its point of attachment. Galls occur singly, in pairs, or rarely, in clusters of four or more. Leaf tissue beyond the point of attachment for these galls usually dies, turning brown. The cup-shaped larval chamber and surrounding flesh measures 3 mm in diameter on top of a 6 to 7 mm long, thin stalk. The stalk is usually brick red to brown, while the cup may be greenish, yellow, or mottled with red. Old galls are usually brown. While some galls are erect, others systematically bend upward (pl. 105), perhaps in response to the orientation of the host leaf. The larval chambers take up most of the space in the cups. There may be some delay with the onset of gall formation to avoid the extremes of summer, or simply two cycles for this wasp: one pulse of gall growth in spring and a second in fall, as both old and fresh galls have been seen in November in the southern Mojave Desert. Adults have been recorded emerging in February and March, with larvae and pupae present later in spring. Little is known about the biology of this wasp. It appears to be a southern species occurring on scrub oaks in the Mojave Desert and other parts of San Bernardino County, as well as on Santa Catalina Island. The gall appeared in Weld's *Cynipid Galls of the Pacific Slope* (1957, fig. 101) and was labeled as unidentified. The species was finally studied and described by Burnett in 1974.

Plate 106. The gall of *Andricus wiltzae* on valley oak.

ROSETTE GALL WASP *Andricus wiltzae*
Pl. 106

This cynipid wasp induces leafy, terminal, polythalamous bud galls on valley oak. These large galls almost always appear at the terminal buds of new stem growth and are recognized by the massive, compact collection of distorted leaves. In spring, the galls are the same green color as normal leaves. They later turn brown. Galls measure up to 40 mm across. Adult females emerge in spring and oviposit in developing leaf buds at the tips of branches. Little else is known about this species. It may have an alternating generation.

PLATE GALL WASP *Liodora pattersonae*
Pl. 107 (syn. *Andricus pattersonae*)

This cynipid wasp induces thin, flat, detachable, monothalamous galls on the underside of the leaves of most white oaks. These summer galls are usually light greenish yellow and measure 7 to 9 mm in diameter and 1 mm thick. The galls have smooth or crenate edges and are narrowly attached under the central larval chamber. The upper surface of the galls is usually smooth, with a central depression marking the larval chamber. Several galls may

Plate 107. Four galls of *Liodora patter-sonae* on the underside of a blue oak leaf.

Plate 108. The galls of *Sphaeroteras trimaculosum* on blue oak.

occur on the same leaf, with the edges of some overlapping with others. Gall development is usually well under way by midsummer. Adults emerge the following spring.

WOOLLYBEAR GALL WASP   *Sphaeroteras trimaculosum*
Pl. 108   **(*Atrusca trimaculosa***
Melika and Abrahamson)

This cynipid wasp induces round-ovoid, monothalamous, solid, bristly galls on the midrib and lateral veins on the underside of valley, scrub, blue, and Oregon oak leaves. These detachable, rusty brown galls are characterized by the stiff, crystalline, erect hairs that uniformly cover the entire surface of the galls. The galls measure 3 to 4 mm across and occur singly or in groups on the leaves. When several of these galls develop in close proximity to each other, they take on the appearance of a woollybear caterpillar. Only parthenogenetic females are known to emerge from these summer galls. Emergence has been recorded at various times of the year, but most exit during late winter and early spring.

Plate 109. The succulent spring galls of *Neuroterus fragilis* on a blue oak petiole and midrib.

Plate 110. The unisexual generation galls of *Neuroterus saltatorius*.

SUCCULENT GALL WASP   *Neuroterus fragilis*

Pl. 109

This cynipid wasp induces integral, polythalamous swellings of the petioles and midrib leaf vein of leather, scrub, and blue oaks. New galls are succulent and green as early as March but turn brown by midsummer. Galls measure 10 to 30 mm long and are most showy on the upper section of the petioles and the lower side of the midrib veins. The swellings severely distort both the petioles and midrib veins, often pinching the base or sides of the leaves together. Males and females emerge from late spring to early summer.

CALIFORNIA JUMPING GALL WASP   *Neuroterus saltatorius*

Pls. 110, 111

This cynipid wasp induces tiny, 1 mm in diameter, round, monothalamous, detachable brown galls on the underside of valley, blue, scrub, and Oregon oak leaves. These late spring unisexual generation galls occur singly, but often in large numbers per leaf. The smooth, glossy, beige brown galls make their first appearance in May and can be found through September. During the

Plate 111. The bisexual generation galls of *N. saltatorius*.

early autumn, the galls begin dropping to the ground by the thousands. The galls get their name from the actions of the larvae once the galls have dropped to the ground. The periodic rapid movements of the larvae cause the tiny galls to flip or jump just enough (about 1 cm) to help work the galls deeper into the leaf material covering the woodland floor. As a result of this action, the galls and larvae may benefit from increased shelter from extreme temperature changes, predation, and other factors that can cause loss. Galls collected in September were still hopping in early November after being refrigerated to simulate natural conditions. An extremely high number of galls are parasitized. Those galls that escape their enemies usually have prepupae in October and November. Females generally emerge within a 20-day span between March and April. The bisexual generation galls develop on leaves shortly thereafter. These integral leaf galls measure 2 mm long by 1 mm wide and are green when fresh, turning brown as they age. This bisexual generation was previously described as a separate species, *N. decipiens.* Rosenthal and Koehler (1971a) found that more than half of the adults of the unisexual generation emerge between 4:45 a.m. and 6:30 a.m., with the remainder exiting throughout the morning. All females seem to leave by 1:30 p.m. They also discovered that none of the adults lived more than 24 hours after emergence. The pteromalid wasp *Guolina psenophaga* is a prime parasite of the summer, unisexual generation gall larvae for this species.

MIDRIB GALL WASP   *Neuroterus washingtonensis*
Pl. 112

This cynipid wasp induces integral, polythalamous galls on the midrib and lateral leaf veins of Oregon oak in spring. The green, abrupt swellings are most noticeable on the underside of the

Plate 112. The midrib gall of *Neuroterus washingtonensis*.

Plate 113. A cluster of the galls of *Trichoteras tubifaciens* on the underside of an Oregon oak leaf.

leaves. The galls turn brown by summer. Galls measure 10 to 30 mm long by 10 to 20 mm wide. Most males and females have emerged by August. Little else is known about this species' biology.

CRYSTALLINE TUBE GALL WASP *Trichoteras tubifaciens*
Pl. 113, Fig. 44   (*Andricus tubifaciens* Melika and Abrahamson)

Figure 44. Cross section of the gall of *Trichoteras tubifaciens* (after Weld 1957).

This cynipid wasp induces unique, tubular-shaped galls with short, crystalline bristles on the underside of Oregon oak leaves. These midrib, monothalamous galls usually occur in tight clusters of up to 35 individual galls. They rarely occur in small groups of a few galls. Clusters measure 20 mm long by 10 mm wide and 10 mm high. Individual tubes are 8 to 10 mm tall by 2 mm wide. The bristly hairs are 1 mm long. The galls vary in color from a solid cream, to yellow with crimson red bristles, to wine red overall. Each gall has a 1.5 mm wide opening at the apex that leads to the top of the

Plate 114. The galls of *Antron douglasii* on valley oak.

thin-walled larval chamber. This feature may allow easy escape of the adults. Females have been found in galls in November, with emergence presumed to occur in winter. Little else is known about the habits of this species.

SPINED TURBAN GALL WASP   *Antron douglasii*
Pls. 114, 115, Fig. 45   *(Cynips douglasii*
     Melika and Abrahamson)

Figure 45. The bisexual generation gall of *A. douglasii.*

This cynipid wasp induces monothalamous, detachable, spiny, unisexual generation galls on the underside of leaves of blue, valley, and scrub oaks. While often flat topped with a stalk or neck, the galls of this wasp can also be round with short spines, or stalkless with spines emerging all over the surface of the gall. The color varies from white with purple-tipped spines to bright pink, rose, or light purple. Some galls have a smoky gray bloom on the surface that rubs off. Galls can measure 15 mm high by 10 mm wide. They are narrowly attached at the base and detach easily. The larval chamber is located near the top of the galls, with an empty chamber below the larva near the base. The function of this cavity is unknown. Females emerge in January and February to oviposit in the buds of the host trees. In spring, much less conspicuous, bisexual generation, monothalamous bud galls develop. The galls of this generation reach a length of 7 mm. Like the galls of the Urchin Gall Wasp *(A. quercusechinus),* the bisexual bud galls of this species are succulent and green, but with longitudinal stripes and tubercles scattered over the sur-

Plate 115. An entire leaf is obscured by several galls of *Antron douglasii.*

face. Males and females emerge in May and oviposit in new leaves. By late June, the unisexual crown galls are well developed. The pteromalid wasp *Quercanus viridigaster* is a parasite of the summer, unisexual larvae.

URCHIN GALL WASP
Pls. 116, 117, Fig. 46

Antron quercusechinus
(Cynips quercusechinus Melika and Abrahamson)

This cynipid wasp induces one of the most spectacular unisexual galls in the western states on blue and scrub oaks. These brilliant, monothalamous galls are either red purple or pink with white tips. They usually occur in groups on the underside of leaves, with several per leaf. The thick, hard spines measure 3 to 4 mm long and often terminate in a point, but some are club tipped or bifurcated. The spines radiate outward across the entire surface of the galls. These summer-fall galls are rather hard, measure to 10 mm across, and resemble sea urchins. The galls either drop attached to leaves in late fall, or break off with the wind and drop to the ground. Adults are found in galls in November and emerge soon thereafter. These parthenogenetic females oviposit in leaf buds. The

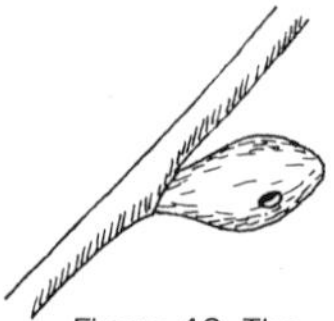

Figure 46. The bisexual generation gall of the *Antron quercusechinus.*

Plate 116. The red purple form of the unisexual generation galls of *Antron quercusechinus*.

Plate 117. The white-tipped pink form of the unisexual galls of *A. quercusechinus*.

monothalamous, bisexual galls that develop as a result are difficult to find, as they are swollen buds that are succulent, thin walled, hollow, green, and ovoid-globular in shape. From these 6 mm long bud galls, males and females emerge. By midsummer, the purple urchinlike galls of the unisexual generation are easily noticed. The pteromalid wasp *Quercanus viridigaster* is a parasite of the unisexual larvae of this species.

CLUB GALL WASP   *Xanthoteras clavuloides* **(syn.**
Pl. 118   *Antron clavula*) (*Atrusca clavuloides*
       Melika and Abrahamson)

This cynipid wasp induces detachable, monothalamous, pubescent, club-shaped leaf galls on valley oak in summer. The galls normally occur singly on veins of the underside of leaves, stand erect, and measure to 8 mm high by 2 mm in diameter. Sometimes, the galls are bent over. The cylindrical galls are characterized by a bulb about three-fourths of the way toward the narrow-pointed tips. The bulge indicates the location of the larval chamber. Most galls have a basal collar at the point of attachment. When fresh and exposed to sun, the galls are green with rose or wine red tones. A single gall disrupts the flow of nutrients enough to starve the outer leaf tissues, resulting in browning and death of sections of the leaves. Females emerge in spring. No alternate generation is currently known.

BALL-TIPPED GALL WASP   *Xanthoteras teres*
Pl. 119   (*Trigonapsis teres*
       Melika and Abrahamson)

This cynipid wasp induces monothalamous, stalked galls on the underside, near the margins of the leaves of Oregon and leather oak. These pubescent, light green, straw yellow, or rose red galls are characterized by a round ball sitting atop a distinctly narrower stalk. With age the galls often turn beige or rusty brown. They almost always stand erect, with several per leaf attached to lateral veins. Galls measure 4 mm in diameter at the bulb by 5 mm high. These galls are usually full grown by July. Unisexual females are present in the galls in November. Emergence and oviposition occurs sometime later. Little else is known about the biology of this species.

HAIR STALK GALL WASP   *Dros pedicellatum* (*Andricus*
Pl. 120   *pedicellatus* **Melika and Abrahamson)**

This cynipid wasp induces thin, horsehairlike, stalked, monothalamous, pointed galls along the margins of blue, valley, scrub, and Oregon oak leaves. A pointed bulge at the tip, which is the larval chamber, characterizes this early spring gall. These galls are

Plate 118 (left). The galls of *Xanthoteres clavuloides* on valley oak.

Plate 119 (right). The galls of *Xanthoteres teres* on Oregon oak.

extensions of leaf veins and measure up to 20 mm high and 2 mm wide at the bulge. The stalk supporting the larval chamber is thin, less than 1 mm in diameter. The spring galls are yellow to orange, glossy, and hairless. By summer, the galls are beige. These smooth galls are fully developed by April, with males and females emerging between April and June. Galls remain on host leaves until fall, even though the occupants have left. This would appear to be the bisexual generation of an as yet undescribed unisexual, summer-fall generation with its own distinct gall.

PEAR GALL WASP
Pl. 121, Fig. 47

Besbicus maculosus (*Cynips maculosus* Melika and Abrahamson)

This cynipid wasp induces monothalamous, pear-shaped galls attached to the midrib vein on the underside of leather and scrub oak leaves. When fresh, galls are green, sometimes mottled, and with a pubescent covering. The narrowest point of the gall is attached to the midrib. At full size, galls measure 9 mm high by 6 mm in diameter at their widest point. With age, a thin outer layer of skin sometimes

Figure 47. The gall of *Besbicus maculosus* showing typical peeling of the outer skin as it ages.

Plate 120. The gall of *Dros pedicellatum.*

Plate 121. The gall of *Besbicus maculosus* on leather oak.

separates from the gall body. Adults emerge in late fall. Fresh galls have been found as late as September, which suggests a long life cycle, different emergence periods, or an alternate generation.

SPECKLED GALL WASP   *Besbicus mirabilis (Cynipis*
Pls. 122, 123   *mirabilis* **Melika and Abrahamson)**

This cynipid wasp induces large, round, hollow, monothalamous galls on the midrib vein on the underside of Oregon oak leaves. Rarely, galls develop on the dorsal surface. These rather spectacular unisexual generation galls are yellow with red spots when fresh in mid- to late summer but eventually turn beige or rusty brown with darker spots. The galls have a velvetlike covering of short and somewhat sparse hairs. Galls measure 25 to 30 mm in diameter. These galls are sometimes called "oak apples" or "pop balls" (for the sound they make when stepped on). They have a thin skin, a central larval chamber, and radiating fibers that support and connect the larval chamber to the outer skin in the otherwise hollow galls. While the reason for this curious form is not yet clear (this form is also repeated among species at high altitude), the air space between the outer gall wall and the larval chamber may provide some temperature buffering effect to protect the larvae from either overheating or freezing (as suggested in the case of the Oak Apple Wasp *[Trichoteras vaccinifoliae]* and the Little Oak Apple Gall Wasp *[T. coquilletti]*). Galls may occur

Plate 122. The gall of *Besbicus mirabilis* on the underside of an Oregon oak leaf.

Plate 123. A cross section of a gall of *B. mirabilis* showing the radial fibers that support the central larval chamber.

singly or in groups of up to three on the underside. When multiple galls are present on the same leaf, they often compress each other, creating flat sides at the points of contact. The galls are fairly resilient, remaining on the leaves during high winds and even after the leaves have fallen. These galls first make their appearance in mid- to late June as small tufts of white hairs on the midrib vein. Young galls are solid until they reach about 3 mm across. As they develop, they become spongy and eventually hollow with radiating fibers and the central, suspended larval chamber. When developing, the skin of these galls is fairly soft and fleshy, but as the galls mature and the larvae complete their growth, the skin loses its moisture and becomes thin and brittle. Parthenogenetic females emerge December through February and lay eggs in leaf buds. Evans (1967) discovered that the larvae from these eggs form the inconspicuous monothalamous bud galls of the bisexual generation. Males and females emerge April to June. Following mating, the females oviposit their eggs into the midrib vein of the leaves, completing the cycle. It has been estimated that only about 4 percent of the larvae of the unisexual generation galls survive to become adults as a result of parasitism. The caterpillar of an inquilinous moth, *Melissopus latiferreanus,* feeds on all of the tissue inside the galls, including the fibers that support the larval chamber, which likely kills the wasp larva. Similarly, I have found Earwigs *(Forficula auricularia)* feeding on the succulent-fleshy fibers

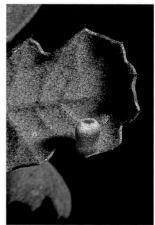

Plate 124. The gall of *Besbicus multipunctatus* on the underside of a blue oak leaf.

Plate 125. The gall of *Phylloteras cupella* on leather oak.

inside fresh galls after having chewed their way through the outer skin in late August. These wasps and their galls are common from northern California through western Oregon and Washington.

GRAY MIDRIB GALL WASP
Pl. 124

Besbicus multipunctatus
(Cynips multipunctatus
Melika and Abrahamson)

This cynipid wasp induces round, pubescent, monothalamous galls on the midrib vein on the underside of blue oak leaves. The light gray pubescence is actually composed of scattered clusters of stellate hairs on top of a darker gray gall surface. Galls measure 8 mm in diameter and are solid and detachable. They usually occur singly, but on rare occasions, two are attached to the same midrib vein. Larval chambers are large, reaching nearly 3 mm in diameter. Adults have been cut out of galls in December.

URN GALL WASP   *Phylloteras cupella*

Pl. 125

This cynipid wasp induces urn-shaped, monothalamous galls on the underside of leaves of scrub, leather, blue, and Engelmann oaks. They usually occur singly and are scattered, but totals of a dozen or more can be found standing erect near the margins of the leaves. The galls have an inrolled lip or edge at the apex. When fresh, the galls are various shades of yellow or red or a combination of both. As they mature, they change color, with the lower half of the gall becoming a dark purple while the top edge becomes rose pink. Old galls turn brown. They measure 2 mm in height by 1.5 mm in diameter, with a central hole forming the urn. These galls are glabrous but have a slightly pitted surface along the sides. Adults emerge in spring. This species was originally reported on Arizona white oak *(Quercus arizonica)*, Rocky Mountain blue oak *(Q. oblongifolia)*, net leaf oak *(Q. reticulata)*, and shrub oak *(Q. subturbinella)* in the Southwest. The galls were also pictured in Weld's *Cynipid Galls of the Pacific Slope* (1957, fig. 135) but were left unidentified at the time because he was unable to separate them from *P. cupella* in the Southwest. It is now clear that the Southwest and California galls are the same species, *P. cupella*. I have found this gall on several occasions in the hills surrounding the Napa Valley and in northern California along the Sacramento River and in Red Bluff and Redding.

Some of the following galls were discovered during the preparation of this guide and are new species. Others were collected by Lewis Weld but left unidentified in his 1957 publication. These cynipid wasp galls have yet to be described but are placed here for the future convenience of the user. (See "About This Guide" in the Introduction.)

PINK BOW TIE GALL WASP   *Undescribed*

Pls. 126, 127

This cynipid wasp induces detachable, monothalamous galls on the underside of blue, scrub, and leather oak leaves. These summer galls develop initially as flat, round discs with a pink center

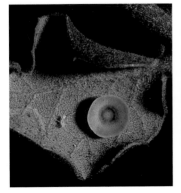
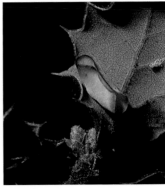

Plate 126. The disc form of the Pink Bow Tie Gall Wasp wasp on blue oak.

Plate 127. The pinched form of the Pink Bow Tie Gall Wasp.

and outer margin and speckled white in between. As the galls mature, the larvae stop feeding, or the galls are attacked by inquilines such as *Synergus* spp., and the galls pinch in from two opposing sides, creating the "bow tie" appearance seen in pl. 127. Galls usually occur on lateral veins and singly or in scattered groups of two or three per leaf. They measure 4 mm in diameter in the disc phase and 7 mm long when pinched in from the sides, suggesting continuing growth after the pinching process. These wasps and their galls are common throughout the range of the host trees. This may be the summer-fall unisexual generation of a species that has a spring bisexual generation, both of which require study and classification. This is a new species that will, hopefully, be described someday.

PINCHED LEAF GALL WASP   *Undescribed*
Pl. 128

This cynipid wasp induces beige pink, monothalamous galls on the underside of the leaves of Oregon oaks. These detachable galls usually occur in large numbers per leaf and appear as though they are pinched inward, like pursed lips. Galls measure 3 mm long by 2 mm wide. The edges of the galls are often dark. Galls are scattered across the undersurface of the leaf between veins and sometimes on veins. Little information is available

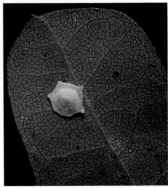

Plate 128. The gall of the Pinched Leaf Gall Wasp on Oregon oak.

Plate 129. The gall of the Disc Gall Wasp on Oregon oak.

on the biology of this species. This gall was pictured in Weld's *Cynipid Galls of the Pacific Slope* (1957, fig. 122).

DISC GALL WASP   *Undescribed*
Pl. 129

This cynipid wasp induces flat, toothy, beige yellow, monothalamous galls on the underside of the leaves of Oregon oak. The galls are distinguished by having randomly occuring, toothlike projections along the outer margins of the galls. These galls occur singly, and rarely more than two per leaf. They measure 4 mm wide and are nearly paper-thin. This gall was pictured in Weld's *Cynipid Galls of the Pacific Slope* (1957, fig. 174).

PINK CONE GALL WASP   *Undescribed*
Pl. 130

This cynipid wasp induces shiny, smooth, pink, cone-shaped, detachable, monothalamous galls with a dark nipple at the apex on the underside of the leaves of Muller's oak in the southern Mojave Desert. It may occur on Muller's and related species of oaks in other desert areas. Galls usually occur one to two per leaf on lateral veins. While the base of the gall is flat bottomed, the sides of the gall rise in an ovoid shape to a point. Galls measure 7 mm

Plate 130. The gall of the Pink Cone Gall Wasp on Muller's oak.

Plate 131. The gall of the Udder Gall Wasp on Muller's oak.

high by 5 mm in diameter. Without study and classification, little else can be said of this unique species. It was discovered in 2004 during the fieldwork for this guide.

UDDER GALL WASP   *Undescribed*
Pl. 131

This cynipid wasp induces oblong, purse-shaped, slightly wrinkled, detachable, monothalamous galls on the underside of the leaves of Muller's oak in the southern Mojave Desert. It may also occur on related species of oaks in other desert areas. These galls usually have a pair of distinctive, prominent, rounded, teatlike projections at the apex. Galls found in November were wrinkled, but it is possible that they lack wrinkles when fresh. These udderlike galls measure 8 to 10 mm high by 7 mm wide. They usually occur one per leaf, but occasionally in separate pairs. This species is common among the oaks in Joshua Trees National Park and was discovered in 2004 during the fieldwork for this guide. Felt (1965) listed a similar looking gall as belonging to *Cynips schulthessae,* but its relationship to this species is unconfirmed.

Plate 132. The gall of the Brown Eye Gall Wasp on Muller's oak.

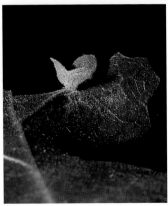

Plate 133. The gall of the Leaf Gall Wasp on valley oak.

BROWN EYE GALL WASP *Undescribed*

Pl. 132

This cynipid wasp induces small, detachable, monothalamous galls with white lateral hairs on the midrib vein on the underside of the leaves of Muller's oak. It may also occur on related species in the southern and eastern Mojave Desert. The apex of the galls is smooth, glabrous, and brown, but the sides of each gall are covered with laterally projecting white hairs, creating the image of "brown eyes." Galls usually occur in clusters of several galls attached to the midrib vein. No single galls have been found. Clusters of four to six galls measure 6 mm wide, while the individual galls in each cluster are only 1 mm high and wide. This is another Mojave Desert species not seen elsewhere and newly discovered in 2004 during the fieldwork for this guide.

LEAF GALL WASP *Undescribed*

Pl. 133

This cynipid wasp induces monothalamous green, leafy extensions of veins on the upper surface of valley oak leaves in spring. These galls tend to project vertically from the surface, mostly along the margins of leaves, and are either tubular or crownlike.

While most galls occur on lateral and marginal veins, some occasionally appear on midrib veins. Galls measure 5 mm high by 3 to 4 mm wide. Individual galls have a dark green upper surface and a lighter lower surface. These spring galls are most likely those of a yet to be determined bisexual generation of perhaps a known unisexual gall maker. Adults have been reared from these galls but have not yet been identified or linked to another generation of a known species.

Intermediate Oak Galls

The galls associated with the intermediate group of oaks, including canyon live oak *(Quercus chrysolepis)*, huckleberry oak *(Q. vaccinifolia)*, Palmer's oak *(Q. palmeri,* syn. *Q. dunnii),* and island oak *(Q. tomentella)*, are distinct from those of the white oak and black oak groups. In some cases, specific cynipid (Cynipidae) genera such as *Heteroecus* spp. are limited to these trees. Strangely, several of the galls that occur on the canyon live oaks of the Pacific states also show up on a variety of canyon live oaks in the Southwest (see table 14 for a list of galls shared by Pacific Coast and Southwest trees). There is also some evidence that Palmer's oak supports a few cynipid wasps not yet found on other members of the intermediate oak group. Unlike the cynipid wasps and the few cecidomyiid midges (Cecidomyiidae) that do not occur in more than one group of oaks, an eriophyid mite (Eriophydae) is found on both black oak and intermediate oak species.

References: Keifer 1952a,b; Weld 1952a,b, 1957, 1960; Lyon 1963; Russo 1979; Dailey and Menke 1980; Keifer et al. 1982; Dailey and Sprenger 1983; Gagné 1989; Hickman 1993.

Stem Galls

Most of the stem galls found within the intermediate oaks group are rather ordinary. This section covers both true stem galls as well as bud galls. In one case you will see a rather interesting feature that seems to facilitate adult emergence from the larval chamber, as decribed below.

TABLE 14 Cynipid Wasp Galls Occurring on Canyon Live Oak *(Quercus chrysolepis)* in Two Distinct Regions

Hairy gall wasp *(Andricus lasius)*

Tapered stem gall wasp *(A. spectabilis)*

Potato gall wasp *(A. truckeensis)*

Split twig gall wasp *(Dryocosmus asymmetricus)*

Golden gall wasp *(Heteroecus melanoderma)*

Intermediate oak disc gall wasp *(Paracraspis guadaloupensis)*

Little oak apple gall wasp *(Trichoteras coquilletti)*

T. frondeum

Little urn wall wasp *(T. rotundula)*

Note: These cynipid wasp species occur in distinct populations of related oaks in the Southwest and Pacific Coast regions. These species were originally listed as occurring on canyon live oak *(Q. chrysolepis)* in the Pacific states and on *Q. wilcoxii* in the Southwest. The latter taxon is now considered a junior synonym of *Q. chrysolepis.*

SPLIT TWIG GALL WASP *Dryocosmus asymmetricus*
Pl. 134

This cynipid wasp induces integral, polythalamous stem galls that rupture and split, exposing a nestlike cluster of larval chambers. This "nest" resembles a clutch of brown, elliptical eggs. The larval chambers are bunched together and not separated by a thick mass of cell tissue as with other polythalamous galls. When fresh, the galls appear as normal stem swellings and measure 5 cm long by 2 cm in diameter. As the galls mature during the summer months, the outer walls on one side of the galls rupture widely, creating large openings into the interior of the galls, exposing the 2 mm long larval chambers. This process converts the once-symmetrical gall into an asymmetrical structure. This lopsided rupturing and congested location of the larval chambers might account for the small amount of damage done to leaves and branches beyond the galls. The flow of nutrients may be able to continue even though some is rerouted into the larval chambers during larval growth. This exposure of the larval chambers at the right time might seem an advantage to the adult wasps, because they only have to chew through the thin walls of the larval chambers, not the thicker, hardened, woody tissue of the gall proper. It

Plate 134. The integral stem gall of *Dryocosmus asymmetricus* showing the exposed larval chambers on canyon live oak.

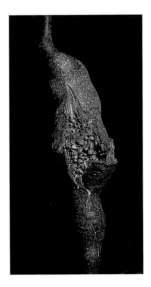

might be, however, a disadvantage due to the exposure of the pupae and adults (prior to emergence) to parasites and predators. Adults appear to emerge before fall. As with some others, this species also occurs on a variety of canyon live oak in the Southwest.

WOOLLY GALL WASP   *Heteroecus dasydactyli*
Figs. 48, 49

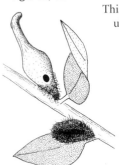

Figure 48. The unisexual generation gall of *Heteroecus dasydactyli* on canyon live oak.

This cynipid wasp induces detachable, globular, monothalamous, woolly, unisexual generation stem galls on canyon live oaks in summer. The galls are covered with a light yellow beige, dense, woolly mass of hairs. The long hairs rub off easily in mats to expose the hard, solid, glossy, spindle-shaped gall beneath. A heavy rain is enough to dislodge the hairs, exposing the gall beneath. The actual gall measures 25 mm in length by 10 mm in diameter. The galls occur singly or in tightly clustered groups. The naked galls bear a ring or collar at the base. The 3 mm thick neck of the galls expands upward toward the bulging main body of the gall, where the larval chamber is located. The bulge tapers toward the apex, which is often recurved. While some galls are straight, others are drastically curved or bent, which may be influenced by inquilines. The surface of the galls is light green to straw yellow and ultimately turns beige with age. There may be scattered tubercles and furrows near the base and apex. Adults emerge near the base of the gall in spring. Females deposit eggs in leaf buds. The bisexual generation galls are monothalamous, integral blister galls that

Figure 49. The bisexual generation gall of *H. dasydactyli.*

appear in the new leaves. Males and females emerge from these galls in late spring to early summer to start the cycle once again.

LYON'S GALL WASP
Figs. 50, 51

Heteroecus lyoni

This cynipid wasp induces nearly flat-topped, monothalamous stem galls with flaring sides known only from Palmer's oaks. The top of each gall can be slightly convex. Specimens from Mexico have ribbed sides and depressed tops. This detachable gall measures about 11 mm high and wide. The sides narrow from the flared top, and the base wraps partially around the stem at the attachment point. The larval chamber is central in the otherwise solid gall.

Figure 50. The typical bud gall form of *Heteroecus lyoni* on canyon live oak.

Galls collected in fall produced adults in late winter and early spring. These are the galls of the unisexual generation. No alternate generation is known. This rather unusual gall and its wasp have been found in select locations on its limited-range host in Riverside, San Benito, and San Luis Obispo Counties, as well as northern Mexico. This species was named in honor of Robert J. Lyon, who contributed so much to cynipid taxonomy and biology. Palmer's oak not only supports a few gall wasps found on other intermediate oaks, but it appears to host at least four species not found on any other oak in this group.

Figure 51. A different form of the bud gall of *H. lyoni,* which may be inquiline influenced.

Plate 135. The gall of *Heteroecus melanoderma*.

Plate 136. The typical unisexual generation bud gall of *Heteroecus pacificus* on huckleberry oak.

The other species unique to Palmer's oak include *Trichoteras burnetti* (bud or petiole gall), Palmer's Oak Stem Gall Wasp (*Loxaulus boharti*, integral stem swelling, described later), and *Heteroecus fragilis* (stem gall with spongy chambers).

GOLDEN GALL WASP   *Heteroecus melanoderma*
Pl. 135

This cynipid wasp induces tubular, detachable, monothalamous, woolly stem galls on canyon live oaks and huckleberry oaks in summer. The short, golden yellow beige hairs rub off only with a persistent effort. The hardened gall beneath has a mealy-granular, nonglossy appearance. These galls do not have a collar at the base but instead rise straight upward, flaring out to form the club shape typical of this species. Galls measure 20 mm in length by 6 mm in diameter. The larval chamber is centrally located. Adults emerge through holes in the neck near the base of the galls. This species also occurs on a variety of canyon live oak in the Southwest.

Plate 137. The bisexual generation galls of *H. pacificus* on the leaves of huckleberry oak.

Plate 138. The unisexual generation galls of *H. pacificus,* probably inquiline or parasite influenced.

BEAKED SPINDLE GALL WASP   *Heteroecus pacificus*

Pls. 136–138

This cynipid wasp induces long, spindle-shaped, monothalamous, glabrous stem galls on canyon live oaks and huckleberry oaks in summer. These unisexual generation galls are often elongated with a slightly curved beak, measuring 30 mm long by 8 mm in diameter. The shape of the unisexual galls can be severely altered by inquilines (pl. 138). Robert Lyon (1963) found inquilinous insects normally associated with the squat and robust forms of this gall, rather than the typical tall, slender varieties. When fresh in July and August, the galls are normally succulent, solid, and mottled green and red or various shades of green and purple, pure green, or beige, with a slight waxy bloom. Fresh galls are usually smooth and glossy, while older galls have wrinkled or cracked surfaces. The elliptical larval chamber is centrally located in a solid mass of tissue. The galls occur singly or in tight groups around the nodes of the stems or terminal buds.

Parthenogenetic females emerge through holes situated near the base of the galls in April and May and oviposit on the underside of new leaves. The resulting bisexual generation galls are conical, thorn-shaped, monothalamous, and often in clusters. These galls measure 6 mm long by 3 mm in diameter. The galls are not easily removed without damaging the leaves. Leaves may be severely deformed where several galls grow closely together.

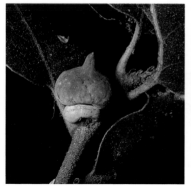

Plate 139. The gall of *Heteroecus sanctae-clarae* on canyon live oak.

Plate 140. The gall of *Disholcaspis chrysolepidis* on canyon live oak.

The bisexual galls are normally wine red to purple. Males and females emerge from the thorn-shaped galls in June. Lyon discovered that only males emerged out of certain clusters of these galls, while other clusters released only females. This suggests that individual unisexual generation females lay eggs that produce either females or males, but not both. Oviposition takes place soon thereafter, producing the summer unisexual generation galls. These galls are common throughout the range of this species' two host trees.

MUSHROOM GALL WASP *Heteroecus sanctaeclarae*
Pl. 139

This cynipid wasp induces monothalamous, detachable bud galls on canyon live oaks and huckleberry oaks that resemble mushrooms or, with a little imagination, Russian church steeples. A round basal section capped with a pyramidal, nippled, detachable top characterizes the two-story gall. The tip of these galls is often slightly recurved. The top half of the galls is usually nonglossy, smooth, or faintly tubercled or furrowed near the apex. The larval chamber is located within the lower unit of the gall where it attaches to the upper half. The entire gall measures 15 mm in height and width. Generally, the two sections of the gall are equal in size, but on occasion, the upper segment is much larger. Adults emerge from the base in April and May. Gall development reaches its peak in July in most areas. No alternate generation is known.

MUFFIN GALL WASP *Disholcaspis chrysolepidis*
Pl. 140

This cynipid wasp induces hard, detachable, humped, polythalamous galls that burst out of the stems in clustered rows on oaks of the intermediate group, including Palmer's oak. When viewed from the side or end, these galls bear the close resemblance to muffins with their overlapping, dark, pitted crust or cap. Fresh specimens have a dark, rusty brown top and light brown sides. Young galls may have a greenish tan color on the sides. The convex top has a cracked, bread-crust texture with a linear ridge, while the sides are relatively smooth. A single gall can measure 15 mm wide by 20 mm long. Clusters can exceed 60 mm in length. The top of a gall flares down from the cockscomblike ridge overhanging the base. The sides narrow downward toward the point of attachment. Each gall can support one or two larvae in separate chambers. When fresh and developing, these galls also secrete a sweet phloem exudate from the base (see the section "Honeydew and Bees, Yellow Jackets, and Ants," in the introduction). Parthenogenetic females emerge through the sides in midwinter. A possible alternate generation has yet to be determined.

TORPEDO GALL WASP *Andricus projectus*
Pl. 141

This cynipid wasp induces bullet-shaped, monothalamous bud galls in the axil of leaves or on catkins in late spring on canyon live oak. When fresh, these solid, elliptical galls are red with a light yellow tip. Some are solid brick red. They measure 10 mm long by 4 mm in diameter. These galls have bud scales around their base. The larval chamber is elongated in the lower half of the gall. The apex of these hard galls is either obtuse or slightly nippled. Females emerge in April either the following spring or as much as three years after gall formation. No alternate generation is known.

Plate 141 (left). The spring bud galls of *Andricus projectus* on canyon live oak.

Plate 142 (right). The honeydew producing gall of *Andricus reniformis* with ants.

KIDNEY STEM GALL WASP
Pl. 142

Andricus reniformis
(*Disholcaspis reniformis*
Melika and Abrahamson)

This cynipid wasp induces detachable, polythalamous, kidney-shaped summer galls on canyon live oaks, huckleberry oaks, and Palmer's oaks. When fresh, the galls are mottled red, yellow, and green, and they produce copious quantities of honeydew, attracting ants (see the section "Honeydew and Bees, Yellow Jackets, and Ants," in the introduction). With age, they turn a tan color. These solid galls measure up to 30 mm long by 15 mm in diameter. The larval chambers are arranged radially around a central core that runs the length of the galls. Parthenogenetic females do not emerge until two or three years after gall formation. Specimens collected in 1973 still contained prepupae in the spring of 1976, with adults emerging a few weeks later. This species is extremely common in the central Sierra, particularly on huckleberry oaks.

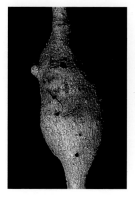

Plate 143. The integral stem gall of *Andricus spectabilis* showing exit holes.

Plate 144. The gall of *A. spectabilis* showing the chew marks of a western gray squirrel exposing the larval chambers to predation.

TAPERED STEM GALL WASP
Pls. 143, 144

Andricus spectabilis
(*Disholcaspis spectabilis*
Melika and Abrahamson)

This cynipid wasp induces integral, abruptly swollen, polythalamous, large stem galls on canyon live oak. The bark-colored swellings measure up to 60 mm long by 30 mm in diameter. The general texture is much like that of normal bark—rough with faint cracks. Damage to the branches and leaves beyond the galls is not common. Gall development begins in late spring and reaches full size by midsummer. Western Gray Squirrels (*Sciurus griseus*) climb out to the edge of branches and chew into these galls, exposing the larval chambers to get at the nutritious larvae and pupae. Those that survive and pupate emerge during the following spring. This wasp also occurs on a variety of canyon live oak in the Southwest. Little else is known about this species.

POTATO GALL WASP
Pl. 145

Andricus truckeensis

This cynipid wasp induces solid, polythalamous, smooth, potato-shaped galls on the stems of canyon live oaks, huckleberry oaks, and Palmer's oaks. At maturity the tan-colored, slightly glossy

Plate 145. The gall of *Andricus truckeensis* showing the typical peeling skin that occurs with age.

Plate 146. The bud galls of *Trichoteras vaccinifoliae* on huckleberry oak.

galls measure up to 60 mm long by 30 mm in diameter. A thin skin that cracks and peels along the edges characterizes this round- to oval-shaped gall. Elsewhere, skin may actually peel off, exposing the pulpy inner tissue of the gall. Normally, no more than a few galls occur on any one branch. Little damage results from the activities of this gall wasp. Parthenogenetic females emerge in January and February in some areas. This wasp is also found galling canyon live oaks in the Southwest.

OAK APPLE WASP   *Trichoteras vaccinifoliae (Andricus*
Pl. 146, Fig. 52   *vaccinifoliae* Melika and Abrahamson)

This cynipid wasp induces round, paper-thin, hollow, monothalamous galls on the stems of canyon live oaks and huckleberry oaks. These summer galls are characterized by being thin, fleshy, and yellow green with red mottling or spots and by their centrally suspended larval chamber supported by numerous radiating fibers connecting it to the outer wall (also see the Speckled Gall Wasp *[Besbicus mirabilis]*). Some galls have noticeable bumps or umbos across the surface. These lightweight galls measure up to 30 mm in diameter. By September most galls have reached their dry, light brown, paper-thin, brittle stage. They usually occur singly or in small groups on older branches, not new spring or

Figure 52. Left: The bud gall of *T. vaccinifoliae* showing the umbos or bumps on the surface. Right: A cross section of the gall of *T. vaccinifoliae* showing the radial fibers that support the central larval chamber.

summer growth. Under the canopy of leaves they are somewhat protected from direct sun. It is possible that the hollow nature and trapped air mass of these galls offer the larvae and prepupae some protection from the freezing temperatures of the snow that covers these trees during winter. This same form is seen in the Little Oak Apple Gall Wasp *(T. coquilletti).* Fresh specimens examined in August usually contain last-instar larvae. Old, dried galls examined at the same time contained prepupae. Specimens with prepupae were placed in refrigeration to simulate winter and removed the following April. Adults emerged within three weeks at ambient temperatures. Adult emergence in the field is probably timed with snowmelt and warming temperatures. This species is extremely common throughout the range of canyon live oaks and huckleberry oaks.

PALMER'S OAK STEM GALL WASP   *Loxaulus boharti*
Pl. 147

This cynipid wasp induces integral, slightly swollen, polythalamous galls known only from Palmer's oaks. These galls are barely recognizable as galls until adults emerge and leave pinhead-sized exit holes. Larval chambers are arranged just under the surface of the bark and measure 2 mm long. Exit holes are about 0.5 mm across with several per linear inch. Usually there is slight swelling around the exit holes, indicating the locations of the galled tissue. Unisexual generation adults emerge in March and April from

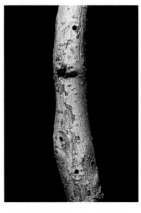

Plate 147. The integral stem galls of *Loxaulus boharti* on Palmer's oak showing exit holes.

two-year-old twigs and from the lower portions of one-year-old twigs. Adults are small, less than 2 mm in length. Eggs are possibly deposited in buds, where the bisexual generation develops. Adults from these bud galls oviposit in twigs in late spring and early summer. This species is restricted to its host in Contra Costa County, Riverside County, and northern Mexico. It may also occur on Palmer's oaks in San Luis Obispo County.

Leaf Galls

A few leaf galls occur on the intermediate species of oaks. For the most part, they pale by comparison with the number of species on white oaks. The organisms that gall the leaves of canyon live oak and huckleberry oak include an eriophyid mite, two cecidomyiids, and several cynipids. As mentioned previously, Palmer's oak supports a unique assemblage of cynipids.

OAK ERINEUM GALL MITE   *Eriophyes mackiei*
Pl. 148

This mite induces erineum galls on the underside of the leaves of oaks in the intermediate group, as well as those in the black oak group. The erineum pockets created on huckleberry oak and canyon live oak are often very deep, with corresponding bumps on the dorsal surface that are more pronounced than with other oaks. The erineum hairs may be white, pink, rose red, or rusty brown. The pockets are often 2 to 4 mm across on huckleberry oak, but can be larger on canyon live oak. This species is unique in using hosts from more than one group of oaks, unlike all of the other insects that gall oaks. Little is known about the biology of this mite.

Plate 148. The erineum galls of the mite *Eriophyes mackiei* on the dorsal and ventral surfaces of a huckleberry oak leaf.

Plate 149. Lower left: Galls of *"Dasineura" sylvestrii* that have joined together on canyon live oak. Lower right: Smaller individual galls of *"D." sylvestrii* on canyon live oak.

FOLD GALL MIDGE   *"Dasineura" sylvestrii*
Pl. 149

This midge induces folded, thickened lobes or pouch galls along the edges of the leaves of canyon live oak and huckleberry oak. It has also been reported on Oregon oak, but I have yet to find it. These galls look more like purses or pouches than a simple fold. Galls show more prominently on the lower surface of the leaves because they are folded in toward the midrib vein without significantly distorting the edge of the leaf, as with other typical fold galls. The galls often appear continuously all along the leaf margin, from the apex to the petiole on both sides, or they are joined to appear as one long gall. Individual roll galls measure 3 mm long by 3 to 4 mm wide. When individual galls occur next to each other, the margin of the leaf has a wrinkled appearance. Fresh galls are lime green and pubescent. Only larvae are known of this species. Gagné (1989) placed this midge provisionally in this genus until adults are reared and the taxonomy becomes clear. This species occurs in California and Oregon.

ROLL GALL MIDGE *Contarinia* sp.
Pl. 150

This midge induces narrow roll galls on the underside of the leaves of both canyon live oak and huckleberry oak. Galls form by rolling the edge of the leaf downward and in toward the midrib vein on the underside. The galls are leaf green. There is usually only one roll gall per leaf, although on occasion you can find both edges rolled. Galls measure 12 to 14 mm long by 1 to 2 mm in diameter. Larvae leave the galls in late summer, although some are found in the galls as late as October. Old galls without the gall midge larvae are brown and usually have exit holes. These gall midges pupate in the leaf litter, and adults emerge the following spring.

PURSE GALL MIDGE Undescribed
Pl. 151

This midge induces rounded, succulent, swollen, monothalamous, smooth-edged fold galls on the underside and along the margins of huckleberry oak leaves. The puffy nature of these galls gives them the appearance of a pouch or purse. They often occur singly or grouped next to or across from each other. I have seen five on one leaf. Unlike the preceding species, the wall of the gall of this species is significantly thicker than the normal leaf. Galls are light green to beige and measure 4 to 5 mm in diameter. The galls of this species are abundant in the central Sierra on huckleberry oak. There is no doubt that these are cecidomyiid galls, based on the larvae, but adults have not been reared to confirm their identification.

Plate 150 (upper left). The galls of *Contarinia* sp. on canyon live oak.

Plate 151 (upper right). The galls of the Purse Gall Midge on huckleberry oak.

Plate 152 (left). The leaf gall of *Heteroecus flavens* on huckleberry oak.

ROUND LEAF GALL WASP *Heteroecus flavens*
Pl. 152

This cynipid wasp induces round, integral, monothalamous galls in the leaves of huckleberry oak. These galls occur singly in the middle or along the margin of leaves, or in groups that can consume the entire host leaf. I have found as many as 10 galls on a single leaf. They appear equally on both sides of the leaves. Galls measure 6 mm in diameter and are either glabrous or bear a short, sparse pubescence. In summer when the galls are fresh, they are green with red mottling on top and light green to yellow on the lower half. Ultimately by fall, they turn a light brown color. The larval chamber is central in a solid mass of tissue. Little is known about the biology of this species.

Plate 153. The gall of the *Heteroecus* sp. on huckleberry oak.

Plate 154. The gall of *Paracraspis guadaloupensis* on canyon live oak.

HAIR CAPSULE GALL WASP

Heteroecus sp.

Pl. 153

This cynipid wasp induces smooth, pointed, monothalamous leaf galls on huckleberry oak. The pointed larval capsules sit atop a hair-thin stalk that emerges out of a lateral vein from the underside of the leaf. Galls occur singly and measure 12 mm long. The larval capsule at the end of the gall measures 1 mm in diameter. Galls are green when fresh but turn beige with age. Little else is known about this species.

HAIRY GALL WASP

Andricus lasius (Disholcaspis lasius Melika and Abrahamson)

Fig. 53

This cynipid wasp induces round, hairy, polythalamous, detachable, midrib vein galls on the underside of canyon live oak leaves. On occasion, these galls will form on large lateral veins. The galls almost always appear a dirty beige color, although when fresh some have golden yellow brown hairs. The hairs are

Figure 53. The hairy gall of *Andricus lasius* on the underside of a canyon live oak leaf.

3 to 4 mm long. The entire gall mass measures up to 10 mm in diameter. Up to six larval chambers are radially arranged near the point of attachment. Females emerge from January to early March. This species also occurs on a variety of canyon live oak in the Southwest.

INTERMEDIATE OAK DISC GALL WASP
Pl. 154

Paracraspis guadaloupensis (*Acraspis guadaloupensis* Melika and Abrahamson)

This cynipid wasp induces convex-flat, disc-shaped, monothalamous galls on the underside of the leaves of canyon live oaks. These beige, spring galls occur singly or up to three per leaf without touching each other. They are usually found on lateral veins. These galls measure up to 6 mm in diameter and are easily damaged and dislodged. Gall development begins by May on last year's leaves. Adults emerge from late winter to early spring. Exit holes are usually off center. While wide ranging, including also occurring on a variety of canyon live oak in the Southwest, this species is not abundant in any one area explored to date.

BOWL GALL WASP
Fig. 54

Paracraspis patelloides (*Acraspis patelloides* Melika and Abrahamson)

This cynipid wasp induces detachable, monothalamous, usually concave galls with flared rims on the midrib or lateral veins on the underside of last year's leaves of canyon live oaks. In late April in the central Sierra the young, fleshy galls are rather flat to slightly concave, but as they mature, the sides of some galls grow higher, creating deep central depressions and forming a bowl shape in some specimens. Other galls remain flat topped. At maturity, these galls measure 12 mm in diameter by 5 mm high. When fresh, the galls are a faint pea green to ivory color with pink red margins. By fall, the galls turn beige with hints of pink or red along the margins of the top. Small dark spots may

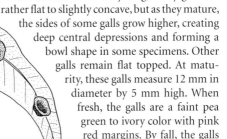

Figure 54. The galls of *Paracraspis patelloides* on canyon live oak.

Plate 155. The leaf gall of *Trichoteras coquilletti* on huckleberry oak.

appear on some specimens. The sides of the galls are ridged and nonglossy. The larval chamber is centrally located at the base of the gall. Adults are found in galls in November and December. Emergence takes place sometime thereafter through holes in the center of the galls. Some emerge as late as May.

LITTLE OAK APPLE GALL WASP   *Trichoteras coquilletti*
Pl. 155   **(*Andricus coquilletti***
                                        Melika and Abrahamson)

This cynipid wasp induces small, round, thin-skinned, monothalamous galls on the underside of the leaves of huckleberry oak and may occur on other oaks in this group. These rather small galls are usually cream yellow with bright red spots and, with age, turn brown. Galls measure 5 to 7 mm in diameter and usually occur singly, but on occasion up to three galls are attached near each other along a vein. Like its larger relative, the gall of the Oak Apple Wasp *(T. vaccinifoliae)*, this gall is hollow with a centrally located and suspended larval chamber supported by radiating fibers that connect to the outer wall. Galls develop from late spring through midsummer. At higher elevations (greater than 1,500 m [4,900 ft]) on huckleberry oak, galls are fresh in July and August. At these elevations, adults probably

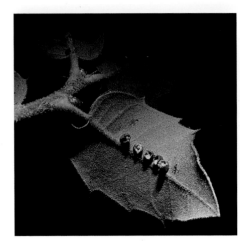

Plate 156. The midrib galls of *Trichoteras rotundula* after the wasps have emerged.

emerge in late spring and early summer. At low elevations (less than 850 m [2,780 ft]) adults emerge during January and February. This species also occurs on a variety of canyon live oak in the Southwest.

LITTLE URN GALL WASP
Pl. 156

Trichoteras rotundula
(*Andricus rotundula*
Melika and Abrahamson)

This cynipid wasp induces small, urn-shaped, monothalamous galls along the midrib vein on the underside of canyon live oak and Palmer's oak leaves. They usually occur in single rows or clusters on both sides of the midrib vein. When fresh, the galls are green with a fine pubescence that ultimately rubs off. With age, they turn brown and smooth. These galls measure 1 mm in diameter and exhibit a small hole at the apex. Branching fibers connected to the outer wall of the gall support the larval chamber. These are the galls of the unisexual generation, without any alternate generation currently known. Adults emerge sometime in fall or winter. This species also occurs on a variety of canyon live oak in the Southwest.

Plate 157. The leaf gall of *Trichoteras* sp. in spring on canyon live oak.

Plate 158. The vein extension gall of the Club Vein Gall Wasp on canyon live oak.

CANYON OAK APPLE GALL WASP   *Trichoteras* sp.
Pl. 157

This cynipid wasp induces round, thin-walled, hollow, monothalamous galls on the underside of the leaves of canyon live oak in spring. As with other oak apples, the galls induced by this wasp have a centrally located and suspended larval chamber supported by radiating fibers that connect it to the outer wall of the gall. These spring galls measure 18 to 20 mm in diameter and are located on the midrib vein. They occur singly or, uncommonly, in pairs. Fresh spring galls are yellow, apple green, lime green, or reddish, with wine red spots scattered across the surface. Galls collected in late July at 1,370 m [4,500 ft] in the central Sierra contained prepupae and were already brown and paper-thin. This spring generation gall species may have an alternate summer-fall generation. This new species was discovered during the preparation of this guide.

CLUB VEIN GALL WASP   Undescribed
Pl. 158

This cynipid wasp induces monothalamous, baseball bat–shaped galls on the leaves of both huckleberry oak and canyon live oak. These large galls are actually extensions of leaf veins that have

Plate 159. The galls of the Starburst Gall Wasp on canyon live oak.

been induced to grow outside the leaf in order to harbor the gall larvae. Galls are usually brick red, brown, or black and measure 7 to 14 mm long by 2 mm at their widest point (the larval chamber). The larval chamber is often half of the length of the gall, unlike that of the Hair Stalk Gall Wasp *(Dros pedicellatum)* on white oaks. Sometimes the sides of the galls are furrowed. Galls often form at the edges of leaves, revealing the source of the vein that has been extended, or directly from a lateral vein. These galls usually occur singly, but sometimes two or three, on either side of the leaf or along the margins. Weld pictured this gall in *Cynipid Galls of the Pacific Slope* (1957, fig. 99).

STARBURST GALL WASP
Undescribed

Pl. 159

This cynipid wasp induces small monothalamous galls along the midrib vein on the underside of canyon live oak leaves in the central Sierra and likely elsewhere. The shape of the galls is unique among gall wasps. Thick spines radiate outward from the central larval chamber, which is covered with bristly, white hairs. When examined under magnification, these hairs appear to be in disarray. The radiating spines can be obtuse, slightly pointed, or joined together, revealing only a toothy edge to the flared base. Galls occur usually in groups of two or three, but with as many as

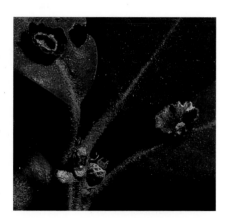

Plate 160.
The galls of the
Funnel Gall Wasp
on canyon live oak.

nine lined up along the midrib vein. Galls are dull pink to ruby red with a lighter-colored larval chamber. Galls measure 2.5 mm in diameter and about 1 mm high. They generally start developing in early summer, but some start as late as early September. Weld pictured this gall in *Cynipid Galls of the Pacific Slope* (1957, fig. 115).

FUNNEL GALL WASP   Undescribed
Pl. 160

This cynipid wasp induces funnel-shaped, monothalamous spangle galls on the upper surface of canyon live oak leaves. Galls are deep purple brown with lines or striations that run from the ragged edge of the galls to the central larval chamber. Galls measure 4 mm wide by 2 mm high. The larval chamber is 2 mm in diameter. The galls flare out and upward from the point of attachment. Galls occur singly, with several leaves in a cluster bearing one gall each. Weld pictured a similar gall in *Cynipid Galls of the Southwest* (1960, fig. 90) but reported it occurring on a white oak. Discovery of these galls in California on canyon live oak suggests the misidentification of the Southwest host oak, or a different gall-inducing species. While it is possible that this is a different species, it is highly unlikely that two galling species with such strikingly similar galls would occur on two different groups of oaks.

Pine Galls

At least 18 species of pine trees are native to California from the coastal dunes to the high Sierra and upper desert regions. Thirteen of these species also occur in other western states. The principal pines considered here include Coulter pine *(Pinus coulteri)*, Jeffrey pine *(P. jeffreyi)*, ponderosa pine *(P. ponderosa)*, gray pine *(P. sabiniana)*, Monterey pine *(P. radiata)*, lodgepole pine *(P. contorta* subsp. *murrayana)*, knobcone pine *(P. attenuata)*, singleleaf pinyon pine *(P. monophylla)*, and Colorado pinyon pine *(P. edulis)*. Most galls associated with pines are best seen on small trees, except where massive witches' brooms are obvious from a distance. Several rust fungi attack and gall pine trees (see table 6). Only one species is described here. Two species of mistletoe that cause witches' brooms are dealt with here, but additional species are listed in table 8. Also included in this section are a few of the gall midges (Cecidomyiidae) that form needle galls.

References: Furniss and Carolin 1977; Sinclair et al. 1987; Gagné 1989; Scharpf 1993; Wood et al. 2003; Miller 2005.

WESTERN GALL	*Endocronartium harknessii*
RUST FUNGUS	*(syn. Peridermium harknessii)*

Pls. 161, 162

This fungus induces globose to pear-shaped, woody stem galls on Coulter pines *(Pinus coulteri)*, Jeffrey pines *(P. jeffreyi)*, ponderosa pines *(P. ponderosa)*, gray pines *(P. sabiniana)*, and Monterey pines *(P. radiata)* in the West, as well as the ornamental Aleppo pine *(P. halepensis)*. This fungus may also appear on other species, including other ornamental pines. Galls occur on stems and exposed roots and continue growing year after year. These galls can exceed 35 cm in diameter. As the galls increase in size, so does the damage to the branches and needles beyond the galls, ultimately killing the branches. During the late spring and early summer months, bright orange yellow aeciospores form in ruptures in the surfaces of the galls and are dispersed each morning for a period of two to three weeks. Heavily infected trees often produce witches' brooms in response to the presence of this rust. Because germination of the spores requires specialized humidity and weather conditions, there appear to be waves of infection that coincide with banner years of highly conducive weather. Unlike

Plate 161. The gall of *Endocronartium harknessii* showing the spores on ponderosa pine.

Plate 162. The witches' broom gall of *E. harknessii*.

other rusts, this one goes from pine to pine without an intermediate host. Some strains of this fungus also pass from pine to paintbrushes (*Castilleja* spp.) Secondary fungi and insects that invade the galled tissues occasionally kill this rust gall. The hyperparasitic fungus *Scytalidium uredinicola* appears to exert some control over the production of spores. The pyralid moth *Dioryctria banksiella* is known to feed on galled tissue (not the fungus) of some pine hosts. This rust fungus has been reported to be limited to North America. Galled trunks and branches wash up on the beaches of Nome, Alaska, having likely drifted down the Yukon River. Trunks and limbs from galled trees are often used as totems and ornate timber for buildings, railings, and fences in Nome and elsewhere. These galls even occur on the stunted shore pines *(P. contorta* subsp. *contorta)* scattered in the muskeg of Alaska.

DWARF MISTLETOES *Arceuthobium campylopodum and A. americanum*
Pl. 163

These mistletoes induce elliptical, integral stem swellings and witches' brooms on Jeffrey pines *(Pinus jeffreyi)*, lodgepole pines *(P. contorta* subsp. *murrayana)*, ponderosa pines *(P. ponderosa)*, knobcone pines *(P. attenuata)*, and Coulter pines *(P. coulteri)*.

Plate 163.
The swollen branch gall of lodgepole pine caused by the dwarf mistletoe *Arceuthobium americanum.*

Galls measure up to 15 cm long by 8 cm in diameter. The swellings are noticeable beneath a burst of succulent, short, bright yellow orange, nearly leafless, but jointed mistletoe stems. Sometimes the elliptical galls are hidden beneath the numerous shoots of the mistletoe or the witches' brooms generated by these mistletoes. As the sticky mistletoe seeds, which are forcibly ejected from their fruits, spread and infection occurs on new branches of the host or other trees, the dispersal of the mistletoe initiates new galls and brooms. *A. campylopodum* is a serious pathogen on Jeffrey pine, while *A. americanum* is common on lodgepole pine.

PINE NEEDLE GALL MIDGE   *Janetiella coloradensis*
Fig. 55

This midge induces needle galls on Colorado pinyon pine *(Pinus edulis),* from Colorado westward. The occurrence in California may be limited to isolated pockets in the mountains of the eastern Mojave Desert. The galls form at the base of the needles and are often covered by the thin, brown sheaths of the fascicle. Galls measure 12 mm long by 4 mm in diameter. Galled needles grow only about one-third their normal length. Color varies from green to purple red and yellow. Galls examined in May bore several 1 mm long, orange larvae in each gall. Larval chambers are enclosed and removable. Larvae do not complete

Figure 55. The needle gall of *Janetiella coloradensis* on pinyon pine.

Plate 164. The needle galls of *Janetiella* sp. on pinyon pine.

Plate 165. The needle galls of *Pinyonia edulicola* on pinyon pine.

their growth during the summer months, but do so the following spring. Adult emergence occurs in late spring, with egg deposition occurring on new needles in May and June. There is only one generation per year. Needles usually turn brown and drop after the adults leave. This species, as well as the next two, do not appear to occur on singleleaf pinyon pine *(P. monophylla)* along the eastern Sierra, but only on Colorado pinyon pine in the West.

LIPPED NEEDLE GALL MIDGE   *Janetiella* sp.
Pl. 164

This midge induces short or stunted, bifurcated needle galls on Colorado pinyon pine *(Pinus edulis),* from Colorado westward, with limited distribution in California in the eastern Mojave Desert. These unique 10 mm long midge galls usually contain two larvae, each in its own depression or chamber. *Contarinia cockerelli,* another gall midge (Cecidomyiidae), has been found in these galls but is not considered the gall maker. The needles do not grow beyond the swelling of the galls. The apex of each gall is characteristically split in two parts. Galls occur side by side near the tips of new growing shoots. Adults emerge through the hole at the apex. The biology of this species is considered to be similar to that of the Pine Needle Gall Midge *(J. coloradensis).*

Plate 166.
The needle galls
of *Thecodiplosis
piniradiatae* on
Monterey pine.

NEEDLE GALL MIDGE   *Pinyonia edulicola*
Pl. 165

This midge induces spindle-shaped galls at the base of Colorado pinyon pine *(Pinus edulis)* needles from Colorado westward, and of single leaf pinyon pine *(P. monophyla)* in the New York Mountains of the eastern Mojave Desert. Eggs are laid on new needles, and the larvae crawl down to the base of host needles where their feeding induces the swollen needle galls. Galls are often light green, sometimes rose red, and measure 8 to 15 mm long by 4 mm in diameter. Galls contain up to 40 larvae that grow during the summer and following spring. Pupation occurs within the gall, with adults emerging early in the second summer. Later, abandoned needles drop off prematurely, before their normal six- to eight-year lifespan.

MONTEREY PINE GALL MIDGE   *Thecodiplosis*
Pl. 166   *piniradiatae*

This midge induces elliptical, spindle-shaped galls at the base of Monterey pine *(Pinus radiata)* needles. This midge has also been reported on Coulter pine *(P. coulteri)* and gray pine *(P. sabiniana)*. Galled needles are usually stunted, about 35 mm long, and do not reach normal size (greater than 110 mm). Actual galls measure 10 mm long by 4 mm in diameter. Normally all three needles within the bundle are swollen. Galls occur among new needles at the tips of shoots in spring. Although the galls bulge out, they are flat at the point of contact with the other needles in the fascicle. Eggs are laid on growing buds usually in January. As with other gall midges on pines, newly hatched larvae crawl to the base of the needles where they begin feeding, initiating the development of the swollen galls. Galls are full grown by February. Pupation occurs sometime in late fall or early winter. Galled needles die in fall.

Redwood False Galls

The tannic acid load in redwoods normally makes them quite resistant to insects. While some fungi attack redwoods under certain conditions, there are no clearly identified gall organisms, even though the following trunk and branch anomalies have been called "galls" by some authors. The following descriptions apply to coast redwood *(Sequoia sempervirens)*. Giant sequoia *(Sequoiadendron giganteum)* may have similar protrusions.

References: Westcott 1971; Scharpf 1993; Scher and Wilson 1996.

Burls

Coast redwoods are famous for large burls around the base of their trunks, as well as monstrously large (greater than 1 m [3 ft]) trunk eruptions on one side of host trees that are quite high off the ground. These lopsided eruptions are most likely burls, since no virus, fungus, bacterium, or insect has been associated with their growth (see the discussion of burls in the introduction). Burl wood is prized for its intriguing grain by furniture and other wood workers.

Branch Tumors

In rare cases, individual trees in the middle of groves of coast redwoods are literally covered with branch tumors (pl. 167) ranging in size from 5 cm to 12 cm in diameter. One such tree exists on the University of California campus in Berkeley. Since no other trees nearby show this malady and no organisms have been associated with their growth, these tumorlike growths are considered genetic anomalies.

Plate 167 (left). Branch tumors on coast redwood.

Plate 168 (right). Cone bract abnormalities of coast redwood.

Cones

Another strange discovery revealed young coast redwood cones with scattered, beige, triangular-shaped bracts protruding sporadically from between cone scales (pl. 168). These bractlike projections measure 3 to 4 mm long. Attempts have been made to isolate a causative agent but to no avail. Further research may reveal a gall organism associated with coast redwoods someday, but for now, there appears to be none.

Plate 169.
The shoot tip gall of *Adelges cooleyi* on spruce.

Spruce Galls

California has three native spruces: Sitka spruce *(Picea sitchensis),* weeping spruce *(P. breweriana),* and Engelmann spruce *(P. engelmannii).* An adelgid induces stem galls on most species of spruce in the West.

Reference: Johnson and Lyon 1991.

COOLEY SPRUCE GALL ADELGID *Adelges cooleyi*
Pl. 169

This adelgid induces integral, elongated, needled stem galls near the tips of new branches in spring on western spruces. This adelgid also galls Colorado blue spruce *(Picea pungens)* and oriental spruce *(P. orientalis).* Galls measure 40 mm long by 20 mm in diameter, with some reaching greater size on occasion. When fresh, the galls are green, fleshy, and covered with normal-sized needles. By midsummer, liplike slits develop above the swollen, expanded needle bases. The adelgids escape these basal chambers through open slits and migrate to the tips of needles, where they turn into winged adult females. These females then migrate to either Douglas-fir *(Pseudotsuga menziesii)* or to another spruce. Eggs laid on the needles of Douglas-fir produce only woolly

Plate 170. Erineum galls of *Aceria erineus* on the dorsal surface of a walnut leaf.

"aphids," not galls. By fall, the galls on spruce look like pineapple-shaped cones, turn brown, and lose the needles. Immature females overwinter on the host trees or their alternate host. They mature in spring as adult stem mothers and start laying hundreds of eggs on the new terminal shoots of spruce trees. This insect is common throughout the West wherever its host trees are found.

Walnut Galls

Two species of walnut are found in California: the English walnut *(Juglans regia)*, and the California black walnut *(J. californica)*. Two gall organisms associated with walnuts are both eriophyid mites (Eriophyidae).

References: Keifer 1952; Keifer et al. 1982.

WALNUT BLISTER MITE   *Aceria erineus*

Pl. 170

This mite induces large, feltlike masses of pale yellow hair on the underside of the leaflets of English walnut *(Juglans regia)*. The hair-lined depressions often measure up to 17 mm in diameter, but occasionally individual erineum galls coalesce to cover much

Plate 171. The galls of *Aceria brachytarsus* on a walnut leaf.

larger sections of the leaflet. There are corresponding lumps that look like blisters on the dorsal surface of the leaflets. Mites overwinter under bud scales and become active when the buds begin to open. This mite is widespread wherever English walnuts grow.

POUCH GALL MITE   *Aceria brachytarsus*
Pl. 171

This mite induces rough-surfaced, veiny, green-red pouch galls on the upper leaflet surfaces of California black walnut *(Juglans californica)*. When abundant, the galls can distort the leaves. The galls occur singly or in coalescing clumps. Galls measure 3 to 8 mm in diameter and stand 3 to 5 mm above the surface of the leaf on short pedicels. The underside of the leaves is marked by corresponding depressions with openings leading up into the galls. These mites overwinter under bud scales. They occur throughout the state.

Willow Galls

California has nearly 30 species of willow (*Salix* spp.), and several more are found in other western states. As with the identification of willows, the gall complex associated with them represents a monumental taxonomic challenge. A colleague quipped, "We would need two more lifetimes to sort out all of the gall insects on willows." In one case, I pursued the identity of a leaf gall organism for over 25 years, finally partially resolving the matter just recently, during the preparation of this guide. Of those we do know about, there are several sawflies (Tenthredinidae), several gall midges (Cecidomyiidae), at least one species of eriophyid mite (Eriophyidae), and a rust fungus (an unknown species not described here) that induce galls on willow.

References: Caltagirone 1964; Smith 1968, 1970; Russo 1979; Keifer et al. 1982; Weis et al. 1988; Gagné 1989.

Stem Galls

At least six common species of sawflies in the genus *Euura* induce integral stem galls on willows in the West, especially California. Unlike other gall organisms, where the larvae initiate gall formation, the adults of this genus, as well as *Pontania* spp. (leaf-gall sawflies), are responsible for gall development. The galls of all *Euura* species reach full development before the eggs hatch. Collateral fluids injected into the meristematic plant tissues by the adults apparently program the willows to produce the galls. Some species of sawflies clean house by ejecting frass from the larval chambers to the outside. During their short, two-week life, the adults of both *Euura* and *Pontania* species drink water, eat pollen and nectar, and browse the pubescence of leaves. In addition to the stem galls induced by these sawflies, there are two other stem galls induced by gall midges of the genus *Rabdophaga*. Keep in mind that the galls described here may occur on willow species other than those mentioned.

POTATO GALL MIDGE   *Rabdophaga salicisbatatus*
Fig. 56

This midge induces large, integral, polythalamous stem galls on several species of willow in the West. The galls are usually abrupt swellings, glabrous, and yellow green when fresh. Occasionally there are minor blemishes on the otherwise smooth surface of the galls. Galls measure up to 40 mm long by 20 mm in diameter. Depending on elevation, these galls usually form in spring on new shoots. Some galls distort the alignment of the affected stems, causing them to bend. These galls can also damage branches and leaves beyond the galls. By fall, the galls are brown and hard. Adults pupate at the surface of the galls in spring. Oviposition follows shortly thereafter.

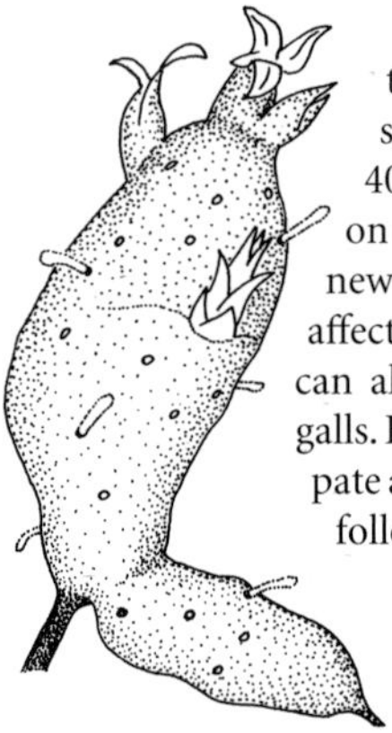

Figure 56. The integral stem gall of *Rabdophaga salicisbatatus* on arroyo willow.

WILLOW ROSETTE GALL MIDGE
Pl. 172, Fig. 57

Rabdophaga salicisbrassicoides

This midge induces open or closed rosette bud galls on several species of willow. These detachable galls with leafy bracts often appear at every axillary joint along the length of the galled stem. They look like old cones when dried and brown in winter. When fresh, however, they look more like lettuce with varying arrangements of leafy bracts. The galls measure up to 20 mm high by 20 mm wide at the base. Leafy bracts often extend away from the main body of the galls and measure 8 mm long by 9 mm wide at the base. Sometimes, the bracts are recurved. In spring, these pliable galls are green, often with red blushing at the base of the bracts. In winter, galls are brittle and often lose most of their leafy bracts. The

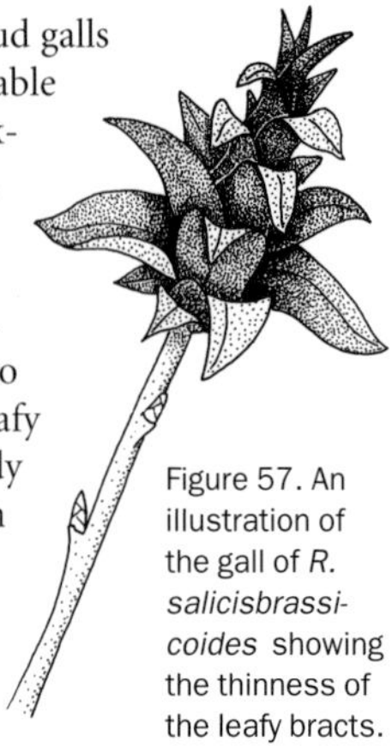

Figure 57. An illustration of the gall of *R. salicisbrassicoides* showing the thinness of the leafy bracts.

larval chambers are located at the base of the galls. Adults emerge in spring and lay eggs in lateral buds. There is only one generation per year. The predaceous midge *Lestodiplosis septemmaculata* has been reared from these galls. There are also several parasitic pteromalid wasps associated with this gall midge.

Plate 172. The cabbagelike bud gall of *Rabdophaga salicisbrassicoides* on arroyo willow.

Plate 173. The gall of *Rabdophaga* sp. showing the thick leafy bracts on willow.

WILLOW CONE GALL MIDGE — *Rabdophaga strobiloides*
Fig. 58

This midge induces pineconelike galls on the lateral and terminal buds of several species of willow throughout the western states. The leafy bracts usually are appressed and not flaring away from the main gall body. When fresh, the galls are green but turn brown with age in fall. Galls usually measure from 10 to 15 mm in length by 10 to 12 mm in diameter. Old galls may remain on the host tree for two or three years. The larval chamber is central near the base. Adults emerge at varying times in spring depending on elevation. There is one generation per year.

Figure 58. The galls of *Rabdophaga strobiloides* on willow.

WILLOW BUD GALL MIDGE — *Rabdophaga* sp.
Pl. 173

This midge induces cone-shaped, monothalamous bud galls on members of the arroyo willow *(Salix lasiolepis)* group. While superficially similar in appearance to the Willow Rosette Gall Midge *(R. salicisbrassicoides)*, the leafy bracts of this species are thick, succulent, strongly recurved, somewhat inflexible, and appressed to the gall body. In between the leafy green gall bracts, the base of the bracts and body of the gall is flushed with red purple. Galls measure to 15 mm long by 10 mm in diameter and develop in summer at elevations above 1,500 m [4900 ft]. Adults emerge the following late spring or early summer.

BREWER'S STEM GALL SAWFLY   *Euura breweriae*
Fig. 59

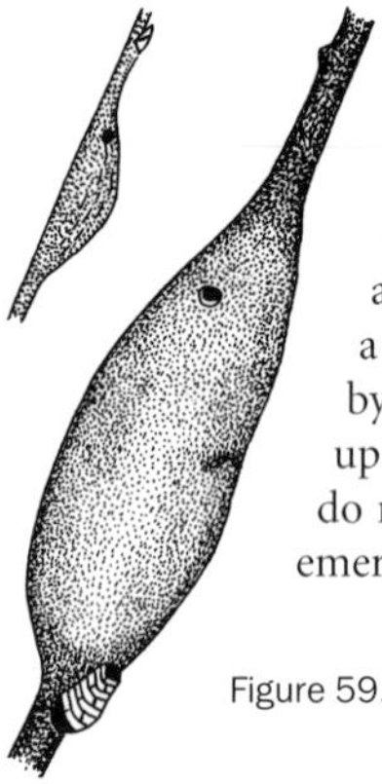

This sawfly induces linear, lopsided, integral stem galls on Brewer's willow *(Salix breweri)*. These glossy, smooth, thin-walled galls develop mostly on one side of the stem. When fresh, these pubescent galls are usually bright yellow green, blushed with a bit of red. Galls measure 20 to 40 mm long by 15 mm in diameter. Each gall can contain up to three larvae. The larvae of this species do not cut exit holes prior to pupation. Adults emerge in April and May.

Figure 59. The gall of *Euura breweriae* on Brewer's willow.

WILLOW STEM SAWFLY   *Euura exiguae*
Fig. 60

This sawfly induces linear, gradually tapering, integral stem galls on narrow-leaved willow *(Salix exigua)* and dusky willow *(S. melanopsis)* and possibly other related species. These smooth-surfaced galls are either with or without minute hairs. At first, the galls are green, but later turn brown as the sawflies mature and stop feeding. Galls measure 20 to 70 mm long by 5 to 15 mm in diameter. The galls are usually lopsided with most of the development occurring on one side of the stem. Galls of this species resemble those of the Arroyo Willow Stem Sawfly *(E. lasiolepis)*. Two or more galls may combine in their development, because of close oviposition, and create what appears as one giant gall. Each

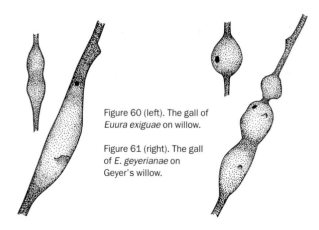

Figure 60 (left). The gall of *Euura exiguae* on willow.

Figure 61 (right). The gall of *E. geyerianae* on Geyer's willow.

gall may contain one or more larvae. When the larvae have completed their development, they crawl to the top of the gall and cut an exit hole, which they plug with frass just prior to pupation. Pupation occurs in the gall. Emergence of males and females occurs in spring as new shoots begin to grow.

GEYER'S STEM GALL SAWFLY   *Euura geyerianae*
Fig. 61

This sawfly induces bulbous, abruptly swollen, integral stem galls on Geyer's willow *(Salix geyerianae)*. Fresh galls are often covered with powdery white material. Mature galls are thick walled and glossy green to mottled red brown. They measure up to 20 mm long by 15 mm in diameter. The larvae make an exit hole prior to pupation in these polythalamous galls. Males and females emerge in late spring. Adults of this species have been observed feeding on nectar, pollen, and the stamens of host willows. This species is abundant in scattered locations from Oregon through the high-elevation conifer forests and stream canyons along the eastern Sierra.

ARROYO WILLOW STEM SAWFLY   *Euura lasiolepis*
Fig. 62

This sawfly induces thick-walled, lopsided, integral stem galls on arroyo willow *(Salix lasiolepis)*. These glossy, glabrous galls occur

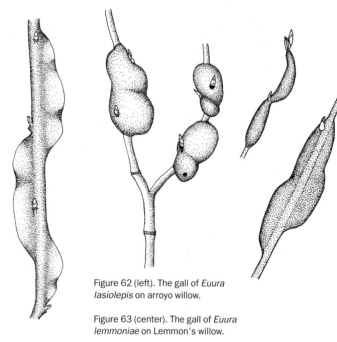

Figure 62 (left). The gall of *Euura lasiolepis* on arroyo willow.

Figure 63 (center). The gall of *Euura lemmoniae* on Lemmon's willow.

Figure 64 (right). The gall of *Euura scoulerianae* on Scouler's willow.

singly on one side of the branch, while some galls develop opposite each other, creating the appearance of one gall. Galls growing in the shade are usually yellow green, while those more exposed to the sun are red purple. These galls measure up to 70 mm long by 20 mm in diameter. The top usually tapers into the stem, while the base is abruptly swollen. As the galls age and dry, they shrivel considerably. Up to five larvae can be found in a single gall. They do not cut exit holes prior to pupation. Precut exit holes increase the potential danger of flooding of the larval chambers, killing the larvae, in some environments. While most species in this genus overwinter in a prepupal state, this species continues to feed and grow slowly until late January. In many parts of the Sierra Nevada, this species pupates and emerges in May, timed with the development of new shoots. Some clones of willows exhibit an inherent resistance to attack by these sawflies, while other clones are susceptible. Several parasitic pteromalid wasps attack this sawfly (as well as other *Euura* species). Galls with walls thicker than the length of the pteromalid ovipositors are less vulnerable to attack by these wasps.

LEMMON'S STEM GALL SAWFLY   *Euura lemmoniae*
Fig. 63

This sawfly induces abruptly swollen, potato-shaped, integral stem galls on Lemmon's willow *(Salix lemmonii)*. Fresh galls in spring are slightly striated, glossy to dull, and dark green to reddish brown. These somewhat knobby galls measure up to 60 mm long by 15 mm in diameter and occur on both sides of the affected stems. Occasionally, some galls will occur in a lopsided fashion on one side of the stem. Normal-looking buds are often found on the sides of these galls. As with other members of the genus *Euura,* the larvae of this species can defend themselves when disturbed by lifting and waving their posterior while ejecting large quantities of pungent fluid from their abdominal glands. Males and females emerge between late April and June. This sawfly is the most common species in high-altitude meadows in the Sierra Nevada.

SCOULER'S STEM GALL SAWFLY   *Euura scoulerianae*
Fig. 64

This sawfly induces thick-walled, elongated, smooth, integral stem galls on Scouler's willow *(Salix scouleriana)*. The galls usually taper toward the top. Galls are usually mottled green and red when fresh, and some are slightly pubescent. With age, they turn brown, then bark gray. Galls measure up to 40 mm long by 15 mm in diameter. The last-instar larvae cut exit holes near the top of the galls and then plug the holes with frass before retreating to the bottom of the galls, where they pupate later. Males and females emerge in May and June. As with the other species in the genus *Euura,* there is only one generation per year. In some areas, as much as 75 percent of the population of this sawfly are killed by parasites. Smith (1970) found that the moisture content of the branches influenced pupation in late spring.

Leaf Galls

At least three groups of organisms are associated with leaf galls that commonly occur on willows: sawflies (Tenthredinidae), gall midges (Cecidomyiidae), and eriophyid mites (Eriophyidae). Also, an unidentified rust fungus galls willow leaves. As with other hosts and gall inducers, the taxonomy of insects on willows is complex and little understood. Several gall midges (Cecidomyiidae) utilize willows but have yet to be identified to species level. I have also found rust, mite, and gall midge galls on willows along the tundra streams of northern Alaska that are undescribed, even though they are abundant and common. Much work needs to be done before we fully comprehend the significance of willows to gall inducers in North America.

WILLOW GALL MITE   *Aculops tetanothrix*
Pls. 174, 175

This mite induces irregularly rounded bead galls on the leaves of several species of willow from Utah and Arizona to California. The warty bead galls measure 2 to 3 mm in diameter and occur singly or in coalesced groups of two or three. The galls appear on both sides of the leaves. Hundreds of these galls can occur on a single leaf, sometimes distorting it. Galls appear between lateral veins and are often yellow green to bright red. Depending on the species of host willow, galls may be glabrous or covered with minute hairs. This eriophyid mite has an alternation of generations with the mites remaining in the galls until fall, when the galls become dry and hard. Mites exit the galls and migrate to crevices in the bark and buds, where they overwinter. Catkins and buds are also galled, but these galls are thought to belong to the eriophyid mite *A. aenigma,* which has also been reported in the eastern United States and in Europe.

WILLOW TOOTH GALL MIDGE   *Iteomyia* sp. A
Pls. 176, 177

This midge induces smooth, large, polythalamous galls mostly on the underside of the leaves of the arroyo willow *(Salix lasiolepis)* and related willows. Each gall is characterized by one or more slightly bent projections that look like the roots of a molar tooth. Each toothy projection is correlated with a larval chamber,

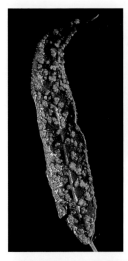

Plate 174 (left). The bead galls of *Aculops tetanothrix* on arroyo willow.

Plate 175 (right). The bud galls of the mite *Aculops aenigma* on arroyo willow.

Plate 176 (left). The gall of *Iteo-myia* sp. A showing several teeth corresponding to larval chambers.

Plate 177 (right). The gall of *Iteo-myia* sp. A showing a single tooth indicating one larval chamber.

allowing you to count the number of larvae per gall. Larval chambers are located near the base of the galls where the galls are attached to the host leaf. Often, a depression surrounds the galls at this juncture where aphids will nestle to feed on cellular liquid. Galls actually bulge on the dorsal surface of the leaves, but the majority of gall growth occurs on the underside of the leaves. The

Plate 178. The galls of *Iteomyia* sp. B. showing the openings at the apices.

galls have no openings, as with the galls of other gall midges. Galls can be yellow green or bright red, depending on exposure to the sun. Galls measure up to 10 mm long by 10 mm across, but single-toothed galls are smaller. Galls collected in August had full-grown larvae. Since gall development does not begin until summer in many locations, the adults must emerge in late spring. Much of the biology of this species remains a mystery. I pursued the identification of this species for many years but was unable to isolate larvae or pupae at the right time of year for identification and rearing purposes until 2004, when I was fortunate to observe galls in the right condition for collecting and later identification.

WILLOW TUBE GALL MIDGE   *Iteomyia* sp. B
Pl. 178

This midge induces round-tubular monothalamous galls that hang from the bottom of arroyo willow *(Salix lasiolepis)* leaves. The distinguishing character of these midge galls is the apical hole or depression that develops at the end of each gall. While most of the gall hangs below the leaf, there is usually a slight depression on the dorsal surface. Galls are commonly pink to rose red with hints of yellow and green. Galls may develop singly or in coalescing groups of four to five individuals. These midge galls measure 5 mm high by 4 mm wide at their base. Gall development begins in spring in the central coast area of California. This species may occur outside of California and throughout the range of the arroyo willow group. While the biology of this

Plate 179. The gall(s) of *Pontania californica* on arroyo willow.

species is relatively unknown, larvae appear to leave the gall by autumn. Presumably, they pupate in spring. Willows that retain their leaves into fall and winter may continue to support the galls of this species.

WILLOW APPLE GALL SAWFLY *Pontania californica*
Pl. 179, Figs. 65–68 **(syn. *P. pacifica*)**

This sawfly induces round-ovoid, monothalamous galls that protrude on both sides of the leaves of related species belonging to the arroyo willow *(Salix lasiolepis)* and Lemmon's willow *(S. lemmonii)* groups. It occurs throughout the West and as far north as the Alaskan tundra. While glabrous and generally smooth, these galls do possess minute, wartlike scales called **LENTICELS** (tubercles). This species is the only sawfly gall inducer in our area whose galls bear such lenticels. The bright green to red galls measure up to 11 mm at their widest dimension. They occur from one to a dozen or more per leaf. Sometimes two or more galls coalesce to form much larger structures. While most galls appear equally on both sides of affected leaves, some galls hang mostly on the underside. Gall formation begins with the injection of collaterial fluids from females. The relatively thin walls that surround the larval chamber are usually 1 to 2 mm thick. Prepupae drop to the ground via silken threads, where they encase themselves in cocoons. Adults usually emerge in spring, timed with the appearance of new leaves. While the typical life cycle involves a single generation per year, under certain environmental conditions

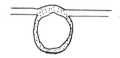

Figure 66. Cross section of a gall typical of the *Pontania viminalis* group (after Smith 1970).

Figure 65. Cross section of the gall typical of the *P. proxima* group (after Smith 1970).

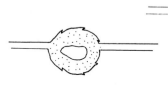

Figure 67. Cross section of a gall typical of the *P. californica* group (after Smith 1970).

Figure 68. Cross section of a gall also typical of the *P. viminalis* group (after Smith 1970).

where leaves remain on the trees throughout the year, up to six generations of this species can occur within a year. Caltagirone (1964) found a remarkable complex of inquiline, parasite, and hyperparasite species associated with this gall maker that included six wasps, a moth, and a weevil. In some cases, it has been found that the moth alone can account for as much as 70 percent of the mortality of the sawfly larvae. There are other similar sawfly galls in our area (particularly in the Sierra) belonging to the *Pontania viminalis* and *P. proxima* groups. The taxonomy and identification of these sawflies and their respective galls in the field is a complicated challenge to say the least.

SHRUB GALLS

Compared to some trees, for example, oaks (*Quercus* spp.) and willows (*Salix* spp.), shrubs generally do not support many galls or gall organisms. Many shrubs have only one gall organism. There are, however, some notable exceptions including creosote bush *(Larrea tridentata)*, rabbitbrush (*Chrysothamnus* spp.), and Great Basin sagebrush *(Artemisia tridentata)*. These widespread shrubs host a fairly large number of gall-inducing insects. Except for roses (*Rosa* spp.), thimbleberry *(Rubus parviflorus)*, and chinquapin (*Chrysolepis* spp.), which host cynipid wasps (Cynipidae), most shrubs support either moths, gall midges (Cecidomyiidae), tephritid fruit flies (Tephritidae), or in one case, a leaf-mining fly (Agromyzidae). In the case of snowberry (*Symphoricarpos* spp.), a sawfly (Tenthredinidae) is the principal gall inducer. Shrubs are listed in alphabetical order of their common names.

Antelope Brush Galls

Antelope brush *(Purshia tridentata),* also called bitterbrush, is widespread throughout western rangeland from the Siskiyou Mountains and Sierra Nevada east. Only two species of gall inducers are common to this shrub. Both are eriophyid mites (Eriophyidae): one occurs on lateral buds, while an undescribed species creates bead galls on the leaves.

Reference: Furniss and Barr 1975; Young and Clements 2002.

ANTELOPE BRUSH BUD GALL MITE   *Aceria kraftella*
Fig. 69
This mite induces round to globular, bristly, rough-surfaced bud galls along current-year stems. These galls usually form at the base of petioles or where axillary buds occur. Galls measure 5 to 12 mm in diameter and are brown or gray like the bark. Galls

often deform the branches, but there is no evidence of stem mortality as a result of this mite. Adult mites emerge in spring to infest new branch growth. This mite and its galls have been found throughout the range of its host, from Montana to British Columbia, Oregon, Nevada, and California. The biology of this mite is not well known.

Figure 69. The nodular gall of the mite *Aceria kraftella* on antelope brush.

ANTELOPE BRUSH BEAD GALL MITE · *Undescribed*
Pl. 180

This mite induces small, round bead galls on the upper surface of antelope brush leaves. The bead galls are light green and measure 1 mm in diameter. Several bead galls can occur on a single lobe of the leaves. No apparent damage appears to result from the feeding of this eriophyid mite. It has been found from Utah to the eastern Sierra.

Bladder Sage Galls

Bladder sage *(Salazaria mexicana)*, also called paper-bag bush, is a common shrub of the creosote bush scrub from the Mojave Desert in California to the Sonoran Desert in New Mexico. Its distinctive papery fruits are easily seen in late summer and early fall when nothing else is left on the naked branches. A gall midge is currently the only known gall inducer on this shrub.

Reference: MacKay 2003.

BLADDER SAGE GALL MIDGE · *Neolasioptera* sp.
Pl. 181

This midge induces tapered, integral, monothalamous stem galls on bladder sage. These symmetrical galls measure up to 14 mm long by 3 mm in diameter. Galls usually occur at joints or on new-growth lateral branches. They are the same color as host stems. Several galls can occur in close proximity to each other, one per branch, and do not appear to coalesce. This species was discov-

Plate 180 (left). The galls of the Antelope Brush Bead Gall Mite.

Plate 181 (below). The gall of *Neolasioptera* sp. on bladder sage.

ered in the Mojave Desert in 2004 during the preparation of this guide. Because this species has yet to be carefully studied and classified, except to identify the genus, there is nothing in the literature, and little else is known about its biology.

Box Thorn Galls

Several species of box thorn (*Lycium* spp.) are found in southern desert areas. One of the more prominent species is Cooper's box thorn *(L. cooperi)*, also called peach thorn, which occurs throughout creosote bush scrub and Joshua tree woodland in the Mojave

Plate 182. The gall of the Stem Gall Moth on box thorn.

and Colorado Deserts of southern California. While the gall described below was found in several locations on Cooper's box thorn, it may also occur on other box thorn species.

References: Miller 2005.

STEM GALL MOTH   *Undescribed*
Pl. 182

This moth, a member of the tribe Gnorimoschemini, induces elliptical, gently tapered, integral, monothalamous stems galls on new spring shoots of Cooper's box thorn. These galls develop when new shoots are green and succulent. New galls are pale green to lavender and measure 30 mm long by 8 mm in diameter. New galls are smooth, glabrous, and sometimes asymmetrical, bulging out on one side of the branch. Old galls harden, become light gray, and may persist on the shrub for several seasons. The larval chamber is fairly large, occupying most of the gall. In April, grayish white moth caterpillars measure 4 mm in length. An adult emerged in November from a gall collected in April. Whether this mimics natural emergence in the field is not known. Little is known about this new species discovered during the fieldwork for this guide.

Plate 183. The bud gall of *Asphondylia enceliae* showing the apical opening on brittlebush.

Brittlebush Galls

Brittlebush (*Encelia* spp.) is a common small shrub on rocky slopes throughout the Mojave Desert and elsewhere. In March, this shrub decorates rocky slopes with its clusters of yellow flowers and is one of the most eye-catching flowering shrubs along major highways in southern deserts. It supports three known gall organisms, with two described here.

References: Gagné 1989; MacKay 2003.

BRITTLEBUSH BUD GALL MIDGE   *Asphondylia enceliae*
Pl. 183

This midge induces white, woolly, globular bud galls along the leafy stems of brittlebush *(Encelia farinosa)* and California brittlebush *(E. californica)*. These galls occur at leaf axils on the main stems and are well developed by mid-March. These galls are much more pubescent than normal stems and leaves. Galls measure 12 mm long by 10 mm in diameter. Each gall has a hair-lined depression at the top where the adults emerge later, usually in February. Little else is known about this species.

Plate 184. The gall of *Rhopalomyia* sp. on brittlebush.

BRITTLEBUSH LEAF GALL MIDGE   *Rhopalomyia* sp.
Pl. 184

This midge induces white pubescent, conical or round, integral, monothalamous leaf, petiole, and flower galls on brittlebush *(Encelia farinosa)* and perhaps other species. Galls can occur on both leaf surfaces. They also protrude on both sides of the leaves, but with most of the galls appearing on one side. These galls are covered with a dense mat of white hairs. The end of each cone-shaped gall has a toothy margin with a hair-lined depression in the center. Galls measure 9 mm high by 6 mm in diameter. There can be from one to six galls per leaf. Galls examined in mid-March had mature larvae or empty pupal cases inside the gall chambers. Since the host bush retains its leaves and adults appear to emerge in spring, this species may have more than one generation per year.

Catclaw Galls

Catclaw *(Acacia greggii)* is a common thorny shrub of the Mojave Desert in California and southern Nevada. This shrub supports several gall-inducing insects including gall midges (Cecidomyiidae) and chalcid wasps (Chalcidae). Unfortunately little is known about their identity and biology. I have recorded at least five species of bud and integral stem galls from Mojave and southern Nevada locations, plus a leaflet gall (figs. 70–72). I am attempting to rear these insects for identification, but sometimes nature reveals its secrets slowly. Three species are described here.

References: Felt 1965; Gagné 1989.

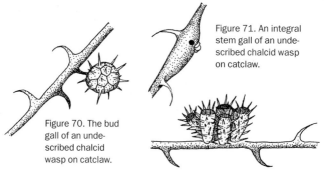

Figure 71. An integral stem gall of an undescribed chalcid wasp on catclaw.

Figure 70. The bud gall of an undescribed chalcid wasp on catclaw.

Figure 72. The tube-shaped bud galls of an undescribed chalcid wasp on catclaw.

CATCLAW LEAFLET GALL MIDGE *Contarinia* sp.
Pl. 185

This midge induces swollen, joined pairs of leaflets on catclaw. Galls are monothalamous. When fresh, galls are green, but because of the relatively short life of the leaflets, the larvae quickly mature and the galls turn brown by mid-March. Galls measure 4 mm long by 2 mm across. These paddle-shaped, glabrous galls have small ridges running their length. These galls occur anywhere on any leaflet, even though most seem to occur on the terminal leaflets. One or two galls can be found on each stemmed cluster of leaflets. Adults emerge in March and April in the eastern Mojave Desert, and timing may vary from one area to another. This relationship is interesting because of the strict timing in the emergence of the adults and the development of the galls during a narrow window when host leaves are available. Most of the year, catclaw remains leafless in response to draught and heat.

ACACIA STEM GALL CHALCID *Tanaostigmodes howardii*
Pl. 186

This wasp induces elliptical, sometimes lopsided, integral, monothalamous stem galls on catclaw. Lopsided galls measure 7 mm long by 5 mm in diameter. Elliptical, tapered galls measure 12 mm long. These galls are the same color as the stems. Galls

Plate 185 (left). The galls of *Contarinia* sp. on catclaw.

Plate 186 (right). The integral stem galls of *Tanaostigmodes howardii* on catclaw.

form on new branches in spring. Adults emerge in the winter. Little else is known about this chalcid.

CATCLAW BUD GALL CHALCID   *Tanaostigmodes* sp.
Pl. 187
This wasp induces detachable, round, monothalamous galls along the stems of catclaw. Galls measure 5 mm across and are gray green, similar to normal stems. Some galls have a barklike texture—gray and cracked. Adults had already emerged from specimens found in mid-December in southern Nevada.

Ceanothus Galls

Over 40 species of ceanothus shrubs (*Ceanothus* spp.) grow in California, along with several additional varieties and subspecies. This widespread group of typically chaparral shrubs is one of the most prominent in the state. For the most part, the only noticeable galls occur on tobacco brush *(C. velutinus),* buck brush *(C. cuneatus),* and related forms. There are three leaf-galling gall midges (Cecidomyiidae), a stem-galling moth (Cosmopterigidae), and a leaf bead–galling eriophyid mite (Eriophyidae).

References: Pritchard 1953: Keifer et al. 1982; Gagné 1989; Johnson and Lyon 1991.

Plate 187 (left). The gall of *Tanaostigmodes* sp. on catclaw.

Plate 188 (right). Bead galls of *Eriophyes ceanothi* on ceanothus.

CEANOTHUS BEAD GALL MITE   *Eriophyes ceanothi*
Pl. 188

This mite induces round-globular bead galls on the undersides of ceanothus leaves, particularly tobacco brush *(Ceanothus velutinus)*. These prominent green galls usually occur in large numbers and can cause distortion of affected leaves. Galls measure 1 mm in diameter unless they coalesce, thus becoming larger. They appear scattered in a random pattern in between veins and along leaf margins. Each gall has a corresponding dimple-depression on the dorsal surface with a hole that allows escape of the adults. The mites and their galls have been found from Colorado to British Columbia and California.

CEANOTHUS STEM GALL MOTH   *Periploca ceanothiella*
Pl. 189   **(family Cosmopterigidae)**

This moth induces integral, spindle-shaped galls on the stems of several species of ceanothus, especially deer brush *(Ceanothus integerrimus)*, in California. Under heavy infestation, this moth and its galls can cause serious damage to branches, stunting the overall growth of the host shrub. As many as 20 galls per branch have been found. Galls measure up to 20 mm long by 5 to 7 mm

Plate 189. Integral stem gall of *Periploca ceanothiella* on ceanothus.

wide and are the same color as normal stems. Larvae overwinter inside their monothalamous galls. They pupate and emerge during spring and early summer. Adult moths lay eggs directly on the stems, and newly hatched larvae burrow into buds and stems. The larvae appear to induce the formation of the galls. There is one generation per year. This species has been reported from New York, Kansas, Texas, Oregon, and California.

CEANOTHUS BUD GALL MIDGE — *Asphondylia ceanothi*
Pl. 190, Fig. 73

This midge induces round, leafy, monothalamous, rosette bud galls on tobacco brush *(Ceanothus velutinus)*. When fresh, these compact bud galls are green to rose red with linear, toothy, pointed leafy bracts, which are arranged hemispherically. The galls measure 12 to 40 mm in diameter. The larval chamber is central. Like other members of the genus *Asphondylia,* this species probably has a fungal associate inside the larval chamber (see the discussion of ambrosia galls in the section on gall midges). After the larvae leave in June and July the galls turn brown. Winter rains hasten the decomposition of the gall's leafy bracts. This species has been recorded in Oregon and California.

Figure 73.
A bract of the gall of *Asphondylia ceanothi* showing the marginal teeth.

Plate 190. The gall of *Asphondylia ceanothi*.

Plate 191. Fold gall of *Asphondylia* sp. on the underside of a ceanothus leaf.

VEIN GALL MIDGE   *Asphondylia* sp.
Pl. 191

This midge induces globular galls on the midrib on the underside of the leaves of tobacco brush *(Ceanothus velutinus)*. These galls have a sparse pubescence and nearly obliterate the midrib. They may superficially resemble the galls of the Midrib Fold Midge *(Contarinia* sp.). Galls measure 7 mm long by 5 mm wide and bulge out more dramatically than the those of the Midrib Fold Midge. There is a corresponding lip or slit on the dorsal surface of the leaves. Larvae leave the galls in July to cocoon in the leaf litter. They presumably emerge the following spring. Little is known about the biology of this gall inducer.

MIDRIB FOLD MIDGE   *Contarinia* sp.
Pl. 192

This midge induces densely pubescent midrib galls on the underside of the leaves of deer brush *(Ceanothus integerrimus)* and tobacco brush *(C. velutinus)* in California and Oregon. Galls often

Plate 192. Fold gall of *Contarinia* sp. on the underside of a ceanothus leaf.

coalesce together to form what appears as a single linear fold. The midrib vein remains prominent along the bottom edge of the galls. Single galls measure 3 to 5 mm in diameter, while clustered galls may reach 15 mm in length. One or more corresponding dimples on the dorsal surface of the leaf mark the pinchpoint along the midrib. Occasionally, the galls appear on main lateral veins. Full-grown larvae leave the galls in July to build cocoons in leaf litter and overwinter until the following spring. There is one generation per year.

Chamise Galls

Chamise *(Adenostoma fasciculatum)* is one of the most common indicator species of chaparral in California. Little attention is usually paid to the insects of such a shrub, yet it plays host to at least three gall-inducing organisms. Only one species, an erio-phyid mite (Eriophyidae), is described here. The other two are a challenge: one, a gall midge, *Asphondylia adenostoma*, attacks and galls seeds without altering their form; the other is an unknown species that induces leafy bract bud galls.

Reference: Keifer et al. 1982.

CHAMISE POUCH GALL MITE   *Phytoptus adenostomae*
Pl. 193

This mite induces globular pouch galls on the leaves of chamise. The galls can severely distort the linear leaves, even though there is rarely more than one gall per leaf. These pouch galls have an ir-

Plate 193. Several bead galls of *Phytoptus adenostomae* on chamise.

regular surface and are usually 1 to 2 mm in diameter. Leaves that are in shade tend to be the ones most frequently attacked by this mite. The galls are a light green, turning brown in age. This mite exhibits an alternation of generations.

Cheesebush Galls

Cheesebush *(Hymenoclea salsola)*, also called burro-bush or winged ragweed, is a common scraggly shrub in washes, sandy flats, and disturbed areas, particularly along roads in the creosote bush scrub, Joshua tree woodland, and shadscale scrub habitats of the Mojave Desert and northwestern Mexico. The peak photosynthetic period for this shrub, like most others in the desert, is in spring when it has new leaves. The gall insects associated with this shrub take advantage of this seasonal growth spurt with the development of their galls in spring and summer. Four gall midges (Cecidomyiidae) are known to gall cheesebush, even though identification to species level has not yet been accomplished. Three midges, *Asphondylia* spp., induce bud galls (two species described here) while a fourth species, *Neolasioptera* sp., induces a tapered stem gall. A fifth gall inducer is a moth belonging to the family Tortricidae, responsible for the stem gall described here.

References: Gagné 1989.

Plate 194 (left). The gall of *Eugnosta* sp. on cheesebush.

Plate 195 (right). Galls of *Asphondylia* sp. A on cheesebush.

CHEESEBUSH STEM GALL MOTH
Pl. 194

Eugnosta sp.
(syn. *Carolella* sp.)

This moth, in the family Tortricidae, induces integral, elliptical, gently tapered, symmetrical, monothalamous stem galls on the upper sections of cheesebush. Galls measure 18 mm long by 8 mm wide. These galls are green when fresh but by midsummer, the stems and galls turn light beige. Galls are striated, with shoots and buds often emerging from the sides. Larvae cut exit holes to the outside, where they later develop the chrysalis and pupate. Most galls are empty in late November, suggesting a fall emergence. The empty chrysalis case is often seen protruding from the exit hole, which is true for all moths in this family. Some specimens collected in November released adults 10 days later. Adults fold their light gray wings back as they rest on stems. Females lay eggs soon after emergence. This new species was discovered during the fieldwork for this guide.

SMOOTH BUD GALL MIDGE
Pl. 195

Asphondylia sp. A

This midge induces green, smooth, glabrous, monothalamous bud galls on cheesebush. These galls occur singly on lateral buds

Plate 196. Hairy bud gall of *Asphondylia* sp. B on cheesebush.

and are **SESSILE**. Some are pointed, while others have an obtuse apex. Galls measure 6 mm long by 2 to 3 mm in diameter. These galls develop rapidly in February and March. By mid-March, adults have emerged from tiny holes in the apex of the galls (pl. 195).

CHEESEBUSH BUD GALL MIDGE   *Asphondylia* sp. B
Pl. 196

This midge induces round, white, woolly, monothalamous bud galls on cheesebush. These fuzzy galls measure 5 to 7 mm in diameter. They are found on axillary buds. Galls occur singly although two or more can occur on a single branch. Adults emerge in May. Egg deposition and gall development occur sometime later. Little else is known about this species.

Chinquapin Galls

Two species of chinquapin, giant chinquapin *(Chrysolepis chrysophylla)* and bush chinquapin *(C. sempervirens)*, occur in coniferous forests and chaparral environments of California. Both appear to support two species of cynipid wasps as gall inducers, one of which has been identified. Both are described below.

Reference: Weld 1957.

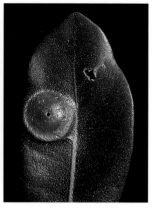

Plate 197. The spring flower gall of *Dryocosmus castanopsidis* on chinquapin.

Plate 198. The gall of the Chinquapin Leaf Gall Wasp.

CHINQUAPIN FLOWER GALL WASP
Pl. 197

Dryocosmus castanopsidis

This wasp induces large, round, monothalamous galls on the flowers of chinquapin. The deep red galls have a golden bloom that easily rubs off. Galls measure 15 mm in diameter. The flesh of the gall is also red with a greenish larval chamber. Galls are easily detached and often litter trails and roadsides. Usually only one gall occurs per staminate inflorescence. These galls have been found in several divergent locations in California and are suspected to occur throughout the range of the host plant.

CHINQUAPIN LEAF GALL WASP
Pl. 198

Undescribed

This wasp induces round, integral, monothalamous leaf galls on chinquapin. Galls are solitary and usually occur equally on both sides of the leaves but in some cases may be more prominent on the underside. They appear to have a raised collar or ring around the base of the gall where the gall emerges from the leaf. The thin-walled galls have a free larval cell (2 mm in diameter) inside a slightly larger cavity (4 mm in diameter). Galls measure

Plate 199. The fold gall of *Sorhagenia nimbosa* on coffeeberry.

11 mm in diameter by 6 mm thick. These spring galls usually occur one per leaf and are green when fresh but turn fawn brown by August. Weld (1957) listed the gall without describing the wasp species.

Coffeeberry Galls

At least two galls are found on California coffeeberry *(Rhamnus californica)* and Sierra coffeeberry *(R. rubra)*, but only one has been identified. The undescribed gall species induces a deformed fruit or berry. Attempts to rear these gall inducers for identification have been unsuccessful. The known gall is that of the moth described below.

MIDRIB GALL MOTH   *Sorhagenia nimbosa*
Pl. 199   **(family Cosmopterigidae)**

This moth induces swollen, monothalamous, midrib galls on coffeeberry leaves. The galls are not classic fold galls as seen with other species. Instead, the galls of this moth are the result of a significant expansion of the tissues in and around the midrib. These rigid, thick-walled galls consume most of the affected leaf with only the basal and apical portions of the leaf unaffected. The sides of the leaf appear pinched together as the swelling consumes much of the leaf. Galls measure 30 mm long by 8 mm in diameter. The galls are usually a lighter green than normal leaf tissues. The single brown caterpillar with a dark brown head is 6 mm long in mid-May along the coast, and later in the Sierra Nevada.

Coyote Brush and Desert Broom Galls

In the United States, 21 species of gall midges (Cecidomyiidae) gall members of the genus *Baccharis*. Coyote brush *(B. pilularis)*, also called chaparral broom, is one of the most interesting shrubs entomologically. Tilden's monumental study (1951) of this shrub identified over 221 species of insects associated with it, as well as eight species of mites. The insects, in turn, hosted an additional 62 species of parasites for a total of 291 species on coyote brush. Coyote brush is a hardy and dominant plant of the coastal chaparral community that exists in central California and extends inland in many areas. Both the prostrate and erect forms of this wind-pollinated shrub host the same gall organisms. There is no evidence of any discrimination between galling agents and male and female plants as host sites for galls. These gall organisms also include a rust fungus, a mite (Eriophyidae), a moth (Gelechiidae), and two gall midges.

A second species of *Baccharis,* called desert broom *(B. sarothroides),* supports a gall-inducing moth, a tephritid fly (Tephritidae), and two gall midges.

References: Tilden 1951; Keifer et al. 1982; Ehler 1987; Gagné 1989, 1995, 2004; Briggs 1993; Headrick and Goeden 1993; Gagné and Boldt 1995; Briggs and Latto 1996; Powell and Povolny 2001.

BACCHARIS RUST GALL FUNGUS　　*Puccinia evadens*
Pl. 200
This fungus induces large, elliptical, swollen stem galls, as well as witches' brooms on coyote brush *(Baccharis pilularis)*. This rust has also been reported on *B. emoryi* and desert broom *(B. sa-*

Plate 200. Stem gall of *Puccinia evadens* showing orange spores on coyote brush.

Plate 201. The pit galls of *Eriophyes bacchariph*a on coyote brush leaves.

rothroides) in California and *B. thesioides* in Arizona. The fissured galls usually reveal the orange uredial spore areas during spring and summer. The galls can exceed 18 cm long by 4 cm in diameter and are usually found on old wood. In the fall, the spore areas are covered with a white powdery material. Through its galling activity this rust fungus disrupts the flow of nutrients and water to the outer sections of branches and leaves, and they often die. This fungus also produces witches' brooms that emanate from the integral stem swellings. The brooms are composed of a compact cluster of small branches up to 20 cm long that often conceal the stem swelling beneath. The brooms usually die after the first season. A gall midge, *Mycodiplosis pritchardi,* feeds on the spores produced in the stem galls by the fungus, with a significant impact on the number of viable spores remaining.

BACCHARIS LEAF BLISTER GALL MITE   *Eriophyes*
Pl. 201   *bacchariph*a

This mite induces swollen blister galls (pit galls) on the leaves of coyote brush *(Baccharis pilularis).* Galls are round to irregular shaped. These greenish yellow, brick red, pimplelike swellings are

1 to 2 mm in diameter and occur on both sides of the leaves. Exit holes usually occur on the underside of the leaves or opposite the main swellings. Each infected leaf may have one or more blister galls. Some leaves are severely distorted. These galls form on new leaves in spring. Little is known of the biology of this mite. A related mite, *Eriophyes baccharices,* causes wartlike, irregularly shaped galls on the leaves of seep-willow *(Baccharis salicifolia).*

BACCHARIS STEM GALL MOTH *Gnorimoschema baccharisella*
Pl. 202

This moth, in the family Gelechiidae, induces hard, integral, round-ellipsoid stem galls on coyote brush *(Baccharis pilularis)* throughout its range. These monothalamous galls are usually well developed as early as June in some areas. Galls measure up to 35 mm long by 15 mm in diameter. They are generally smooth, green, and hollow. While some galls are more elliptical, tapering gradually, others are abrupt swellings. Gall development begins after overwintering eggs hatch in the spring and the larvae burrow into new terminal shoots. The occupied stems swell around the larvae, leaving large cavities within which the larvae feed. In April in coastal counties, larvae are usually 5 to 7 mm long. They nearly double their size to 10 to 12 mm in length by July. By June or July, however, many of the moth larvae have been parasitized or lost to other predators. Usually from late May until mid-July surviving larvae cut exit holes through the gall walls and drop to the ground to pupate. Adults emerge August and September. Eggs are laid in fall on the outer branches. The frass that accumulates inside the gall serves as a culture medium for various fungi, which are eaten by several fungus insects after departure of the moth. Tilden (1951) found at least 10 parasites associated with this moth in addition to 17 other insects that were connected to the gall or the moth in some manner. The intricacies of these complex relationships stagger the mind for such a common, yet so disregarded, shrub.

DESERT BROOM GALL MOTH *Gnorimoschema* sp.
Pl. 203

This moth, in the family Gelechiidae, induces terminal, mono-thalamous leaf galls on desert broom *(Baccharis sarothroides)* in

Plate 202. The integral stem gall of *Gnorimoschema baccharisella* on coyote brush.

Plate 203. Terminal leaf gall of *Gnorimoschema* sp. on desert broom.

the northern Mojave Desert and perhaps throughout the range of the host. Galls are composed of tightly bound, thick, swollen leaves that are held together like the staves of a barrel, similar to the galls on rabbitbrush (*Chrysothamnus* spp.). (Because these leaves are swollen and larger than normal leaves, I consider them galls.) These dark green galls measure 40 mm long by 7 mm in diameter and are twisted at the apex. The larval chamber is large, nearly the full length of the gall. By mid-March there are 4 mm long whitish caterpillars in the galls with frass packed at the basal end. I have seen shrubs near Red Rock Canyon State Park with nearly every terminal shoot galled. Pupation occurs outside the galls.

DESERT BROOM GALLFLY   *Aciurina thoracica*
Fig. 74

This fly induces club-shaped to elliptical, integral, monothalamous stem galls on desert broom *(Baccharis sarothroides)* in southern California. Galls are usually smooth and yellow green when young. As the galls mature and reach full size, they develop red streaks and ultimately turn brown. Vegetative shoots and flower stalks will project out of the galls. Fully developed galls

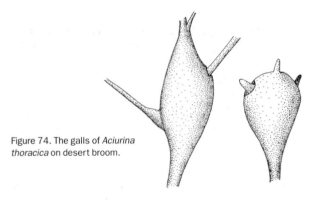

Figure 74. The galls of *Aciurina thoracica* on desert broom.

become somewhat corky and rough with age. Eggs are laid near terminal and axillary buds in the early spring. Fly larvae feed mostly on sap from parenchyma cells. Adults emerge from January to March and live from 60 to 85 days. Late season rains may induce a second growth spurt among host plants and sometimes a possible second generation. There is some evidence that an alternate host, coyote brush *(B. pilularis)*, along the Central California coast, may be used by a second generation. For the most part, however, this species is considered to have a single generation per year.

COYOTE BRUSH BUD GALL MIDGE
Pl. 204

Rhopalomyia californica

This midge induces globular, polythalamous terminal bud galls on both the erect and prostrate forms of coyote brush *(Baccharis pilularis)*. The galls are round to lobed with lumpy protrusions across the surface. The soft, smooth, spongy, green to red purple galls measure up to 20 mm in diameter. Galls often have leaves emerging from them. Females lay clusters of eggs on terminal buds. While on the exterior of the host, the eggs are often heavily parasitized by the wasp *Platygaster californica*. Eggs that survive unparasitized produce larvae (generally in spring) that burrow between bud scales and commence feeding. The gall tissue swells around each of the larvae. Complete development of this midge can occur in as little as 30 days, but often as much as 70 days or more. When fully grown, larvae burrow to the surface of the galls, where they develop their partially protruding white cocoons

Plate 204. Terminal bud galls of *Rhopalomyia californica* on coyote brush.

and pupate. This species represents one of the rare situations among gall insects where fresh galls and emergence of adults occur throughout the year, depending on location and environmental circumstances, even though there is a pulse of growth and gall activity in spring. I have found prostrate forms of this shrub along the sand dunes of California's central coast where every terminal bud had been galled. This gall midge is certainly one of the most common and abundant gall midges in California.

Studies have found seven parasitic wasps associated with this gall including the species mentioned above. Competition for hosts among parasites seems to impact total parasitism effectiveness on the host. In combination, some parasitism effectiveness rates exceeded 80 percent loss of the gall midge. Parasite influence appears to have a major impact on population control for this midge species. It has been introduced in Australia and Texas as a biological control for nonnative *Baccharis* spp. As mentioned earlier, the intricacies of host plants, gall inducers, parasites, and inquiline relationships, especially as exemplified by the coyote brush regime, are vastly complicated and a worthy subject for additional study in the future.

TUBE GALL MIDGE   *Rhopalomyia sulcata*
Pls. 205, 206

This midge induces columnar, glabrous, monothalamous leaf galls on several species of *Baccharis* but especially desert broom (*B. sarothroides*) in the Mojave Desert in California. These blunt-tipped galls are usually found growing at an oblique angle out of the middle of the leaf or from the tip. There is usually some swelling or broadening of the leaf at the point of emergence.

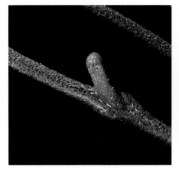

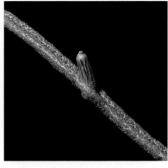

Plate 205. The leaf gall of *Rhopalomyia sulcata* on desert broom.

Plate 206. Another color form of the gall of *R. sulcata* on desert broom.

Shallow furrows mark the surface of the galls. Galls measure 3 mm high by 1 mm in diameter. Color varies from solid green to beige with vertical purple stripes. These midges have been reported galling various species of *Baccharis* from the southwestern United States (Texas, New Mexico, and Arizona) and Mexico.

COYOTE BRUSH STEM GALL MIDGE
Pl. 207, Figs. 75, 76

Rhopalomyia baccharis

This midge induces integral, twisted and bent, striated, polythalamous stem galls initially at the tips of new shoots of coyote brush *(Baccharis pilularis)* in spring (a fall crop of galls also develops). By midsummer, fast-growing spring shoots have elongated well beyond the location of the early spring galls. Galls are green in the spring but look similar to older bark by midsummer. Mature galls measure 50 to 90 mm long by 5 to 8 mm in diameter. Each linear gall can contain several individual larval chambers, which are usually located just below each of the bends in the gall. A single gall can bend several times. These galls stand out from the normal straight shoots as a result of this twisting and bent

Figure 75. Illustration of the stem gall of *Rhopalomyia baccharis*.

Plate 207. Integral stem gall of *Rhopalomyia baccharis* on coyote brush.

appearance. Larval chambers are linear, 5 to 10 mm long by 1 to 2 mm wide, and located just inward of the cambium layer of the stems. In some areas, fresh galls with orange larvae can be found in fall, suggesting that this midge has two or more generations per year. Based on this, it is presumed that adults from spring galls emerge in late summer and oviposit in the slower-growing branch tips, where fresh, green, fall galls are found through winter. Studies during fall have shown that the galls develop elliptical holes to the outside at the top of the larval chambers and just above branch nodes. The rounded-edged, elliptical holes are not typical of the normal sharp-edged, round exit holes created by insects, and they do not appear to have been chewed open. While we have no clear answer at this point, the convenient exit holes appear to be created by the plant after larvae have stopped feeding and stimulation from the larvae has ceased. These holes develop before the larvae (at the bottom of the chambers) change into pupae. Shortly after the holes are created, pupae develop and remain at the bottom of the chambers.

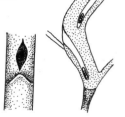

Figure 76. Cross section of the stem gall of *R. baccharis* showing the positioning of the larval chambers and pupae with preexisting exit holes (note the atypical shape of the exit hole).

Pupation occurs inside the galls, and the adults emerge through the openings in November and December. These unique circumstances make this species stand out from all of its relatives in which the pupae partially push their way out of the gall before the adult emerges. Much more needs to be learned about the habits of this mysterious and interesting gall midge.

TEARDROP GALL MIDGE   *Rhopalomyia* sp.
Pl. 208

This midge induces round, pointed, monothalamous leaf galls in the spring on desert broom *(Baccharis sarothroides)*. These colorful galls are yellow, pink, or wine red with a thin, sharp point at the apex, giving them the appearance of small teardrops. Galls measure 3 mm high by 2 mm wide with a slightly granular surface. These tiny galls emerge from the leaf at an oblique angle. They may be found on the same branches and leaves as the Tube Gall Midge *(R. sulcata)*. Normally, however, there is usually one gall on a leaf. This midge may be found throughout the range of its host. Little else is known about this new species discovered during the fieldwork for this guide.

Creosote Bush Galls

Creosote bush *(Larrea tridentata)* is the dominant shrub in much of the desert regions of southern California, Nevada, Utah, Arizona, Texas, and Mexico. When driving through these desert areas, you can see endless miles of nearly pure creosote bush. Creosote bush is remarkable for its habit of growing outwardly from the parent plant, producing clonal rings genetically identical to the original parent, some of which measure up to 9 m (30 ft) across. One of these giant "fairy ring" structures has been dated as up to 11,700 years old, which if accurate may make creosote bush the oldest living plant on earth. Surprisingly enough, creosote bush is the host to at least 16 species of gall midges (Cecidomyiidae) and, therefore, is a critical host plant for gall insects, including their parasites and inquilines. Nearly all of the gall midges known to date are from the genus *Asphondylia,* whose members are also known to have symbiotic relationships with fungi (see the discussion of ambrosia galls in the section on gall

Plate 208 (left). Leaf gall of *Rhopalomyia* sp. on desert broom.

Plate 209 (right). Bud gall of *Asphondylia apicata* on creosote bush, showing pupal case.

midges). There is also a single species of *Contarinia* that induces a common gall, but the adults have yet to be reared for species determination. One critical study of creosote bush indicated a higher number of galls and gall-inducing species on plants growing in nutrient- or water-stressed circumstances than on those host plants that were significantly less stressed (see the section "Environmental Factors"). The galls of 15 of the 16 known species are described here. These galls and their associated midges are widespread, occurring throughout the range of the host plant.

References: Waring 1986; Gagné 1989; Waring and Price 1989, 1990; Gagné and Waring 1990; Phillips and Comus 2000.

CREOSOTE BUSH GALL MIDGE *Asphondylia apicata*
Pl. 209

This midge induces monothalamous bud galls with either a straight or slightly bent projection emanating from the apex of the main gall body. Galls are green when fresh but turn a chocolate brown with age and have yellow or pale orange projections. Galls measure 7 mm long overall, with the lower swollen segment about 4 mm long by 2 mm in diameter. The thin projection is usually 3 mm long and less than 1 mm in width. Galls are usually found near the terminal buds of creosote. Larvae and pupae have been found both in spring (March) and early autumn (Septem-

Plate 210. The rosette bud galls of *Asphondylia auripila* on creosote bush.

Plate 211. Cluster of leaf galls of *Asphondylia barbata* on creosote bush.

ber), and in November I have found galls with pupae. The season of adult emergence may be stretched out over a long period of time, or there may be two generations per year for this gall midge. This species has been found from Arizona to Death Valley and the southern Mojave Desert.

CREOSOTE STEM GALL MIDGE   *Asphondylia auripila*
Pl. 210

This midge induces round, leafy-bract, polythalamous galls on the stems of creosote bush. The golf-ball-sized galls can be easily seen from the road and measure up to 25 mm in diameter. They are green when fresh. The galls are composed of hundreds of thin, 10 mm long, green or brown bracts that are arranged in multiple, small rosette patterns. The bracts arise from each of the individual club-shaped to pyramidal galls comprising the larger gall mass. These individual galls measure 7 mm high by 3 mm wide at the base. Fresh green galls can be found almost year-round, but most likely following major rain events. Each individual gall within the larger golfball-sized cluster has a terminal plug through which the larvae emerge to build their white cocoon at the surface. Adults emerge from spring to late summer, although there may be some variation in different regions. Over 17 species

Plate 212. Club-shaped leaf gall of *Asphondylia clavata* showing the hinged exit hole lid.

of parasites have been associated with *Asphondylia* spp. galls. Nine species apparently account for 98 percent of the parasitism. Research has shown that parasites attack the Creosote Stem Gall Midge all through the stages of its development. The Seri Indians of the Sonoran Desert in the Southwest smoked the dried galls like tobacco.

CREOSOTE BUSH MIDGE *Asphondylia barbata*
Pl. 211

This midge induces small, elliptical, ovoid, slightly pubescent, nondetachable, monothalamous galls, often in clusters, on the underside of leaves of creosote bush. Galls are usually green, sometimes with a dark brown spot near one end. Galls measure 2 mm long by 1 mm in diameter. They are often found side by side on the small leaves. Larvae and pupae have been found in spring and early fall. Adult emergence is presumably in spring, but the biology of this and other creosote bush species remains elusive.

CREOSOTE BUSH CLUB GALL MIDGE *Asphondylia clavata*
Pl. 212

This midge induces glabrous, club-shaped, monothalamous galls on the leaves of creosote bush. These galls resemble those of the

Plate 213. The galls of *Asphondylia digitata* on creosote bush.

Plate 214. Leaf gall of *Asphondylia discalis* showing the exit hole on creosote bush.

Leaf Club Gall Midge *(A. pilosa)* except for the absence of hair. The galls of this species often terminate in a more pronounced swollen larval chamber than those of the Leaf Club Gall Midge. Galls measure 4 mm long by 1 mm in diameter at the widest point. They occur singly and usually extend off the midrib on the underside of the leaf. Occasionally, you can find both species' galls growing on the same leaf. As with other *Asphondylia* species, larvae and pupae have been found over a wide period in spring and early fall.

CREOSOTE ANTLER GALL MIDGE   *Asphondylia digitata*
Pl. 213

This midge induces monothalamous, antlerlike leaf galls on creosote bush. These green, leaflike galls have noticeable toothy margins and look like moose antlers. Galls measure 5 mm across and are the thickness of normal leaves. As with the normal paired leaflets, two galls often form close together. Fresh larvae-containing galls have been collected in January. Pupation occurs inside the galls, with adults emerging later. Females deposit eggs directly into leaf tissues with their short ovipositors.

CREOSOTE BUSH PADDLE GALL MIDGE Asphondylia discalis
Pl. 214

This midge induces paddle-shaped, monothalamous galls that usually grow at right angles to the host leaves. When fresh, the galls are green but turn dark brown with age. Galls arise from the midrib on the underside of the leaves. Galls measure 4 mm long by 1.5 mm wide. They are flat and leaflike and occur singly or in pairs. Larval chambers are located in the center of the paddle-shaped galls. Larvae and pupae have been found at several times of the year from spring through fall. This species may have more than one generation per year.

CREOSOTE BUSH GALL MIDGE Asphondylia fabalis
Pl. 215

This midge induces stringbean-shaped, monothalamous galls that protrude from the underside of the leaves. These galls occur singly or sometimes two or three per leaf. Galls measure up to 5 mm long by 1.5 mm in diameter. They are somewhat rounded in cross section, not laterally flattened as in some related galls. Some galls are slightly bent. Galls have a hairless, slightly mealy-granular surface and are green to brown. Larvae and pupae have been found in spring and fall. As with other members of the genus *Asphondylia,* the larvae have three instars, are usually white to yellow, and pupate inside the galls with pupal cases usually protruding from the galls.

CREOSOTE BUSH FLOWER GALL MIDGE Asphondylia florea
Pl. 216

This midge induces club-shaped galls at the base of the flowers and fruiting bodies of creosote bush. These galls are easily missed, as they superficially resemble the hairy seed pods of the bush. Galls are 4 mm long by 1 mm in diameter and are covered with long white hairs, just like the seed pods. Larvae and pupae have been collected mostly in spring. Pupae emerge through holes at the apex of the galls. Galls examined in November were empty. The biology of this species is not well known. It has been found from Arizona to the southern Mojave Desert.

Plate 215 (right). Gall of *Asphondylia fabalis* on creosote bush.

Plate 216 (lower left). The seed gall of *Asphondylia florea* showing exit hole.

Plate 217 (lower right). The gall of *Asphondylia foliosa* on creosote bush.

LEAFY BUD GALL MIDGE   *Asphondylia foliosa*
Pl. 217

This midge induces round, monothalamous, singular stem galls. These galls do not occur in clusters, as with other species. Most galls in January are brown. When fresh, presumably in spring, the galls consist of numerous slender, green, leafy bracts without completely covering the base of the gall body. The leafy bracts tend to be rather short, with the entire gall measuring 7 mm in diameter. Pupation occurs inside the gall.

LEAF CLUB GALL MIDGE   *Asphondylia pilosa*
Pl. 218

This midge induces monothalamous, club-shaped galls that stand erect on the leaves of creosote bush. These finely pubescent

Plate 218. Galls of *Asphondylia pilosa* on a leaflet of creosote bush.

Plate 219. Bud gall of *Asphondylia resinosa* on creosote bush.

galls occur singly or in pairs. Galls measure 7 mm high by 1.5 mm in diameter at the widest point—the larval chamber. Galls are beige near the base and dark gray on the upper section. A similar species, the Creosote Bush Club Gall Midge *(A. clavata)*, has galls that are smooth and glabrous, otherwise they are difficult to distinguish.

CREOSOTE RESIN GALL MIDGE   *Asphondylia resinosa*
Pl. 219

This midge induces small, round, monothalamous stem galls. These galls are usually covered with a hard, glossy brown resin. Some galls are round with appressed leafy bracts around the base. Other variants of this gall have protruding leafy bracts, arranged in clusters of four. Galls measure 3 mm in diameter and are usually an olive green color. These galls have been collected fresh in January. Little is known about the life history of these rather large midges.

CREOSOTE CONE GALL MIDGE   *Asphondylia rosetta*
Pl. 220

This midge induces leafy, monothalamous, cone-shaped galls on stems. These galls are distinguished from the Leafy Bud Gall

Plate 220. The bud gall of *Asphondylia rosetta* on creosote bush.

Midge *(A. foliosa)* by the recurved bracts that cover the entire gall body. The bracts have a sparse pubescence over their surface. Galls are green when fresh and measure 10 mm high by 4 to 5 mm wide. They are brown in winter. Larvae have been collected in August and September, with pupae collected as late as April. Pupation occurs in the galls in spring. Emergence and reproduction occur shortly thereafter.

LEAF POD GALL MIDGE   *Asphondylia silicula*
Pl. 221

This midge induces podlike, pointed, laterally flattened, monothalamous galls on the underside of the leaves of creosote bush. The galls hang from the middle of the leaf or from along the margins. One or more galls can occur per leaf. Galls measure 3 mm long by 1 to 2 mm wide and are green to brown. As with other members of this genus, larvae and pupae can be found from spring through fall.

SCIMITAR LEAF GALL MIDGE   *Asphondylia* sp.
Pl. 222

This midge induces flat-sided, sword-shaped, striated, monothalamous bud galls on creosote bush. These galls are green when fresh but turn brown with age. The laterally flattened sides are furrowed with shallow ridges. The apex of the galls is usually obtuse, and the whole gall is arched. These distinct galls are almost always found on terminal buds and stand out from normal leaves. Adults have not been reared for identification to species level. Their behavior is likely similar to that of other *Asphondylia* adults.

Plate 221 (left). The gall of *Asphondylia silicula* on creosote bush.

Plate 222 (lower left). Bud gall of *Asphondylia* sp. on creosote bush.

Plate 223 (lower right). The gall of *Contarinia* sp. on creosote bush.

CLASPING LEAF GALL MIDGE   *Contarinia* sp.
Pl. 223

This midge induces galls composed of two leaves joined together to create flat, circular galls. These monothalamous galls are green and look almost like normal leaves except for their broader, swollen nature. The galls actually look like green wings on terminal branches. Galls collected in January were either green or brown, indicating either multiple generations per year or a delay in the development of some of the galls. They measure 8 mm high by 6 mm wide and occur in pairs or clusters of several. This species, yet to be determined, appears to be the only insect to gall creosote bush that is not a member of the genus *Asphondylia*. Little else is known about this midge, since adults have not been reared and studied.

Plate 224. Dorsal view of the pocket gall caused by *Exobasidium vaccinii* on false azalea.

Plate 225. A ventral view of *E. vaccinii* showing the extent of the bulge underneath.

False Azalea Galls

False azalea *(Menziesia ferruginea),* also called mock azalea, is a common shade-tolerant forest shrub along the northern coast of California, extending north to Alaska. It is host to a fungus that severely deforms the infected leaves.

References: Westcott 1971; Sinclair et al. 1987; Pojar and Mac-Kinnon 1994.

LEAF GALL FUNGUS *Exobasidium vaccinii*
Pls. 224, 225

This fungus produces swollen, fleshy, white, bladderlike pockets on leaves of false azalea in the spring. Besides the swollen leaves, the fungus can also induce stem galls and witches' brooms. The fungus also attacks buds and and flowers. The leaf galls are usually bowl- or cup-shaped bags consuming part or all of the leaf. Galls measure up to 25 mm long by 13 mm wide. Leaf galls bulge out on the lower surface, with a corresponding deep depression or pocket occuring on the dorsal surface. Galls found in June were sporulating from the exterior surface, giving the galls a powdery white appearance. Spores produced in summer overwinter among bud scales until the following spring. This fungus is found throughout the United States and Canada. It has also been recorded on west-

Plate 226. Integral stem gall of *Neolasioptera* sp. on woolly sunflower.

ern azalea *(Rhododendron occidentale)*, bearberry *(Arctostaphylos uva-ursi)*, blueberry *(Vaccinium* spp.), cranberry *(Vaccinium macrocarpon)*, gaultheria *(Gaultheria* spp.), huckleberry *(Vaccinium* spp.), bog laurel *(Kalmia microphylla* subsp. *occidentalis)*, madrone *(Arbutus menziesii)*, manzanita *(Arctostaphylos* spp.), bog rosemary *(Andromeda polifolia)*, rhododendron *(Rhododendron* spp.), and Labrador tea *(Ledum groenlandicum)*, among others. It has been reported that Native Americans along the British Columbia coast picked and ate these fungus galls.

Golden Yarrow Galls

Golden yarrow *(Eriophyllum confertiflorum)*, also called woolly sunflower or false heather, is common along the north and central coasts. This small shrub hosts a stem-galling midge. No other gall-inducing insects are known at this time to occur on this host plant.

Reference: Gagné 1989.

STEM GALL MIDGE   *Neolasioptera* **sp.**
Pl. 226

This midge induces an integral, round to elliptical, monothalamous stem gall near the tips of golden yarrow *(E. confertiflorum)*. The galls are covered with a fine, white pubescence, as are the stems and leaves of the host plant. Some galls are abrupt swellings, while others are slightly tapered. Round galls measure 5 to 7 mm in diameter, while the tapered versions are usually 10 mm long by 7 mm across. The inner flesh of the galls is green. The large larval chambers are lined with a white fungus (see the discussion of ambrosia galls in the section on gall midges). Orange larvae have been found in galls in May, even though some had already exited the galls through holes near the top. The life history of this midge is not well understood. It may also occur on the related seaside woolly sunflower *(E. staechadifolium),* which shares similar habitats. It was initially found in 2003 in Morro Bay during the preparation of this guide.

Goldenbush Galls

Several species of goldenbush (*Ericameria* spp.) occur in the West. Goldenbush *(E. ericoides)* is a common shrub inhabitant of coastal sage scrub communities. Its relative, cliff goldenbush *(E. cuneata* var. *spathulata),* is common among rock outcroppings in the Mojave Desert. Goldenbushes likely support several species of gall-inducing gall midges (Cecidoyiidae), three of which are described here.

References: Gagné 1986, 1989.

ROSETTE BUD GALL MIDGE   *Rhopalomyia ericameriae*
Pl. 227

This midge induces leafy, rosette bud galls with flaring and slightly recurved bracts on Palmer's goldenbush *(Ericameria palmeri)* and goldenbush *(E. ericoides).* These galls measure 7 mm long by 5 mm wide and are green when fresh. At first the leafy bracts remain erect, but they recurve and turn brown with age. Bracts are 1 to 2 mm wide. Larvae develop during spring and early summer. These midges pupate inside the galls. Adults emerge in April to oviposit in new buds. Little else is known about the biology of this species.

GOLDENBUSH GALL MIDGE   *Rhopalomyia* sp.
Pl. 228, Fig. 77

This midge induces a knobby, rosettelike, sticky, polythalamous, terminal bud gall on cliff goldenbush *(Ericameria cuneata* var.

Plate 227 (left). The gall of *Rhopalomyia erica-meriae* on goldenbush.

Plate 228 (right). The bud gall of *Rhopalomyia* sp. on desert goldenbush.

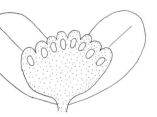

Figure 77. Cross section of the terminal bud gall of *Rhopalomyia* sp. showing the position of the larval chambers.

spathulata). This bud gall has numerous blunt projections that extend from the top of the gall. Each knob seems to be associated with a larval chamber immediately below. Larval chambers are arranged radially near the surface of the gall. Galls measure 7 to 10 mm high and wide. Fresh galls in late November are round and extremely sticky. I have not found old galls, so I do not know what they look like with age. On some shrubs, nearly every terminal bud is galled. Presumably the adults emerge in spring. This species occurs in the Mojave Desert and perhaps elsewhere in the range of its host. It was discovered in 2004 during the preparation of this guide.

GOLDENBUSH BUD GALL MIDGE   *Prodiplosis falcata*
Pl. 229

This midge induces round to slightly elongated, cone-shaped, monothalamous bud galls on goldenbush *(Ericameria ericoides).* Some elongated galls measure 10 mm long by 7 mm wide, but

Plate 229. The gall of *Prodiplosis falcata* on coastal goldenbush.

most old, brown galls are 8 mm by 8 mm. Bracts measure 5 to 8 mm wide at the base. Fresh galls in May are green with broad leafy bracts with outwardly pointed tips. A fine, white pubescence shows along the edges of each bract. Old galls are brown and look more like closed cones, with the bracts appressed. The pointed tips of the bracts apparently break off with age. This species is particularly prominent in the dunes of the San Luis Obispo coastline.

Gooseberry Galls

Gooseberry (*Ribes* spp.) occurs in a variety of habitats. Galled species found to date have generally occurred in open forest environments. The gall inducers associated with these shrubs include an aphid (Aphididae) and a gall midge (Cecidomyiidae).

References: Gagné 1989; Johnson and Lyon 1991.

GOOSEBERRY LEAF-ROLLING APHID   *Macrosiphum*
Pl. 230   *euphorbiae*
This aphid induces swollen, leaf-roll galls on some gooseberries. These roll galls can completely obliterate terminal leaves, rolling both edges in toward the middle. The veiny galls appear lighter green than normal leaves. This aphid is also known as the Potato Aphid, as it infects numerous vegetable crops with many alternate hosts in the summer including gooseberry, honeysuckle (*Lonicera* spp.), and crabapple (*Malus* spp.). This aphid is common throughout the United States and Canada.

Plate 230. Roll galls of *Macrosiphum euphorbiae* on gooseberry.

Plate 231. Old gall of *Rhopalomyia* sp. on gooseberry.

BUD GALL MIDGE   *Rhopalomyia* **sp.**
Pl. 231

This midge induces green, leafy, monothalamous bud galls in the spring. These galls are composed of numerous, thin leafy bracts that flare out in an open rosette pattern and competely conceal the gall body. Galls measure 10 to 12 mm across. These are spring galls that turn brown by August, indicating a cessation of larval activity. Adults probably emerge in spring as new buds swell. Little is known about the biology of this midge.

Gumweed Galls

Gumweed (*Grindelia* spp.) is a common plant in a variety of habitats. Little attention has been paid to this genus as a host to gall inducers. However, at least three gall inducers warrant attention here: a moth (Gelechiidae) and two midges (Cecidomyiidae).

References: Gagné 1989; Powell and Povolny 2001.

GUMWEED GALL MOTH   *Gnorimoschema grindeliae*
Pl. 232

This moth, in the family Gelechiidae, induces large, integral, monothalamous stem galls only on the lower branches of *Grin-*

Plate 232. Integral stem gall of *Gnorimoschema grindeliae* on gumweed.

Plate 233. The gall of *Rhopalomyia grindeliae* on gumweed.

delia hirsutula. Galls usually occur close to the ground and not in the outer branches. Sometimes galls cause a clustering of branches above the galls. Galls are ovoid and measure 20 to 22 mm long by 12 to 14 mm in diameter. They are composed of overlapping leaf bracts with one or more normal leaves protruding from the gall body. Most grow in the shade, although a few receive sunlight. Galls are usually green, but those exposed to sun may be red purple. These thick-walled, succulent galls have rather large larval chambers up to 14 mm long. Larvae develop rapidly, reaching full size in April. Most larvae cut an exit hole near the top of the gall and drop to the ground by the end of April. Larvae create large cocoons up to 15 to 30 mm long in the sand and leaf litter under their host plants. This is a host-specific moth whose life cycle is keyed to the habits of its host. These perennial plants usually die back by the end of summer, and it is presumed that the adult moths deposit their eggs near the base of the plants, where they may overwinter.

FLOWER GALL MIDGE *Rhopalomyia grindeliae*
Pl. 233

This midge induces sticky, modified, capsulelike, hard, mono-thalamous galls among the flowers of *Grindelia hirsutula*. This

Plate 234. Leafy rosette gall of the Gumweed Gall Midge on gumweed.

gall may appear on other gumweed hosts in California and adjoining states. Galled florets usually stand out from those unaffected by the midge. Galls measure 5 mm long and have a hole at the apex. These galls are green when fresh in early spring. Flower heads with several galls can exceed 25 mm across. Adults emerge and lay their eggs in October. Since the host plant dies back by fall, the eggs must overwinter. Little else is known about the biology of this species.

GUMWEED GALL MIDGE   *Undescribed*
Pl. 234

This midge induces large, green, polythalamous, leafy bud galls on *Grindelia hirsutula*. These rosette galls are composed of numerous, thin, long, green, leafy bracts. The overall size of the galls can reach 40 mm across, with leafy bracts that measure 1 mm wide by 25 mm long. These galls form in spring on terminal and lateral buds as they begin to swell. Because the host plants tend to die back in fall, it is presumed that the adults emerge in fall. This new species was discovered in central California during the fieldwork for this guide.

Plate 235. The gall (center) of the Hazelnut Bud Gall Midge.

Hazelnut Galls

Hazelnut *(Corylus cornuta)* is usually found under the canopy of bay-oak woodland or other shady, moist habitats. In the West, there appears to be a single gall-inducing midge associated with hazelnut, although the gall inducer is still not known.

References: Felt 1965; Gagné 1989.

HAZELNUT BUD GALL MIDGE *Undescribed*

Pl. 235

This midge induces round to globular, monothalamous bud galls on this host. The larvae of both a *Contarinia* species and a *Dasineura* species have been associated with these galls. The galls are actually aborted buds that never develop or open. During the growing season in spring, the galls are 7 to 10 mm in diameter and a normal bud green color. By fall, the buds turn brown and dry. At this time the larvae have ceased feeding and enter diapause. Adults have yet to be reared, which would help to determine the species. It is presumed that they emerge in spring, timed with the development of spring buds. This bud gall is common in California. Little else is known about the biology of this insect.

Honeysuckle Galls

Several species of honeysuckle (*Lonicera* spp.) occur in the West, including introduced ornamental varieties. One of the more common native species is *Lonicera subspicata,* which occurs in dry habitats and chaparral slopes along the coast from central to southern California. It supports two distinct bud galls—both induced by gall midges (Cecidomyiidae). A second honeysuckle common in the central western part of the state, *L. hispidula,* supports a distinct bud gall made by another gall midge. These galls are induced by three closely related species, only one of which has been described. The other two may ultimately be described as new species belonging to a new genus, which will likely include the one species currently known. The relationship between the taxonomy of these three midges needs further study before we can get a clearer picture of the honeysuckle gall community.

Reference: Gagné 1989.

HONEYSUCKLE BUD GALL MIDGE   *Rhopalomyia lonicera*
Pl. 236

This midge induces round to globular, fleshy, succulent, monothalamous bud galls on *Lonicera subspicata.* These spring galls are green and measure 10 to 15 mm in diameter. Bud scales often flare out at the tips of the smooth galls. Larvae overwinter and pupate inside the galls, with adults emerging February to March. Little else is known about the biology of this midge.

LEAFY ROSETTE GALL MIDGE   *"Rhopalomyia"* sp. A
Pl. 237

This midge induces cabbagelike, terminal, monothalamous bud galls on *Lonicera subspicata* and possibly other species. The tips of galled shoots have from one to four galls clustered together, making them look like one large gall. The leafy bracts covering the exterior of the galls are large, broad bud scales to 7 mm wide. Galls are leaf green and measure 14 mm wide by 5 mm high. Larval chambers are large, reaching 3 mm in diameter. The inner wall is smooth and bright green except for a small white patch of fungal material at one end of the chamber. Pupae begin to

Plate 236 (above). The gall of *Rhopalomyia lonicera* on honeysuckle.

Plate 237 (left). Terminal bud gall of *"Rhopalomyia"* sp. A on honeysuckle.

develop in January along the central coast of California. This gall inducer is placed in the genus *Rhopalomyia* provisionally, until such time as the proper genus and species is assigned. Adults have been reared from these galls for later taxonomic determination. This new species was discovered during the fieldwork for this guide.

COMPACT BUD GALL MIDGE　　　*"Rhopalomyia"* sp. B
Pl. 238

This midge induces conical, leafy, monothalamous bud galls on *Lonicera hispidula*. While these galls can occur singly, they usually appear in congested clusters that can reach 30 mm across. Individual galls measure 10 mm high by 8 mm in diameter, including the leafy bracts. The bracts of these galls are much more linear and narrower (3 mm wide) than those of the Leafy Rosette Gall Midge (*"Rhopalomyia"* sp. A) (7 mm wide). Long, white hairs protrude from the edges of each bract. Also, each gall has a dense arrangement of white hairs that fill the center of the galls, marking the entrance to the larval chamber. Larval chambers measure 4 mm high by 2 mm wide. In February, galls appear fresh and green. Those exposed to direct sunlight often turn wine red. Half

Plate 238. The gall of *"Rhopalomyia"* sp. B on honeysuckle.

of the galls examined in late February contained larvae, while the other half contained pupae. Adults emerge in February and March. Females are relatively large midges that have a dark brown thorax and orange abdomen. Males have smaller, thinner abdomens. It is not clear when galls begin development, given the apparent freshness of galls in winter. This gall inducer is placed in the genus *Rhopalomyia* provisionally until such time as the proper genus and species is assigned.

Horsebrush Galls

Several species of horsebrush (*Tetradymia* spp.), also known as cotton-thorn, occur in the West. Mojave horsebrush (*T. stenolepis*) is one of the more prominent species common through the Joshua tree woodland and creosote bush scrub communities and supports a stem-gall-inducing moth (described here). A coastal species, *T. comosa,* supports a woolly, ovoid, lateral bud gall induced by the gall midge "*Mayetiola*" *tetradymia* (not described here).

References: Hartman 1984; Powell and Povolny 2001.

HORSEBRUSH STEM GALL MOTH   ***Scrobipalpopsis***
Pl. 239   ***tetradymiella***
This moth, in the family Gelechiidae, induces white, woolly, integral, globoid, monothalamous stem galls on cotton-thorn (*Tetradymia axillaris*), *T. glabrata,* and Mojave horsebrush (*T. stenolepis*). Galls generally occur at the base of new shoots in spring. Fresh galls often have leaves and spines emerging from the sides of the galls. Old galls may remain on the shrub for two or three seasons. These galls are symmetrically tapered toward the tip, from which the shoot continues. These galled shoots rarely

extend more than 7 cm in length, far short of normal shoots (greater than 14 cm). Galls measure to 15 mm long by 8 mm in diameter. They may vary in size depending on the host species. The walls of the galls tend to be thin (1 mm), as the larval chamber is quite large, occupying most of the available space for larval feeding and growth. A 15 mm long gall had a 13 mm long larval chamber, for example. Galls initiated prior to normal and rapid shoot elongation tended to be smaller than those initiated during bursts of spring growth. One study revealed that galls were more abundant on plants growing at lower elevations in the Mojave Desert. Plants on the edges of washes or roads seem to have more galls than those competing with other desert shrubs for limited water that are farther away. Pupation occurs within the galls. Adults emerge during June and July, are gray mottled, and measure 8 to 10 mm long. They remain active primarily during the morning hours through mid-August. A second generation of larvae feed from late June to mid-September, with the adults emerging in late September. Egg deposition occurs through November or until the onset of cold weather. Overwintering eggs hatch in April, with stem gall development continuing through June.

Huckleberry Galls

Blueberry, huckleberry, and cranberry belong to the genus *Vaccinium*. Five species found in California extend all the way into southeast Alaska. These include dwarf blueberry *(V. caespitosum)*, black huckleberry *(V. membranaceum)*, evergreen huckleberry *(V. ovatum)*, red huckleberry *(V. parvifolium)*, and bog blueberry *(V. uliginosum)*. Three types of galls occur on various species of huckleberry, including those induced by a fungus, an unknown moth, and another gall-inducing species currently unknown to entomologists.

References: Westcott 1971; Sinclair et al. 1987; Johnson and Lyon 1991.

HUCKLEBERRY BROOM RUST FUNGUS   *Pucciniastrum*
Pl. 240   *goeppertianum*
This fungus induces witches' brooms on evergreen huckleberry *(Vaccinium ovatum)*, dward blueberry *(V. caespitosum)*, red huck-

Plate 239. Shoot gall of *Scrobipalpopsis tetradymiella* on horsebrush.

Plate 240. Witches' broom caused by *Pucciniastrum goeppertianum* on huckleberry.

leberry *(V. parvifolium),* and black huckleberry *(V. membrana-ceum),* among others in western states. The stems comprising the witches' brooms are glossy, swollen, and mahogany brown and twist and turn in a compact cluster. Several clusters of brooms can occur on a single shrub. These affected branches die after the growing season. Galled branches are usually 15 to 17 cm long and bear leaves. The fungus is systemic and perennial. The telial stage of the fungus is responsible for the broom galls on blueberries and huckleberries. This fungus has no **UREDINIAL STAGE,** as do some other rusts. This rust has been called the fir-blueberry rust be-cause the **AECIAL STAGE** of the rust uses fir trees (subalpine fir *[Abies lasiocarpa],* grand fir *[A. grandis],* noble fir *[A. procera],* white fir *[A. concolor],* red fir *[A. magnifica],* and balsam fir *[A. balsamea]*) as alternate hosts. Basidiospores produced on the brooms infect fir needles, allowing the development of the aecial stage. This rust fungus is found in the northern United States, in-cluding nearly all of the western mountains from Alaska south to Mexico.

Plate 241. The gall of the Leaf-rolling Moth on huckleberry.

Plate 242. Bud gall of an undescribed organism on huckleberry.

LEAF-ROLLING MOTH   *Undescribed*
Pl. 241

This moth induces roll galls along the edges of huckleberry leaves, especially *Vaccinium membranaceum*. Roll galls usually range from 10 to 15 mm long by 2 mm in width. These galls are green like the host leaves. Larvae develop through late spring and summer. Adults may emerge in late summer to deposit their eggs on the branches for overwintering. There is one generation per year.

HUCKLEBERRY BUD GALL ORGANISM   *Undescribed*
Pl. 242

This organism induces red, succulent, pea-sized, monothala-mous bud galls at the base of the petiole or where the petiole and leaf join on *Vaccinium ovalifolium,* and perhaps other species as well. Galls generally show the bud tip protruding from the swollen gall itself. They sometimes consume the entire petiole. These bud galls measure 5 to 10 mm in length by 5 mm in diameter and are usually red on the side exposed to sunlight and yellow green on the shady side (pl. 242). Several unsuccessful attempts have been made to rear the gall inducer.

Plate 243. Nitrogen-fixing nodules caused by the bacterium *Rhizobium* sp. on lupine found on the relatively barren outwash plain of the Baird Glacier in southeastern Alaska.

Lupine Galls

Over 70 species of lupine (*Lupinus* spp.) occur in California, and several additional species are endemic to other western states. Of all the species examined, two coastal species appear regularly to support midge galls, yellow bush lupine *(L. arboreus)* and blue bush lupine *(L. chamissonis)*. Three gall midges and a nitrogen-fixing bacterium are described here. Gall-inducing bacteria are common to all lupines, as well as several other plants (table 5).

References: Sinclair et al. 1987; Gagné 1989; Gagné and Strong 1993.

NITROGEN-FIXING BACTERIUM   *Rhizobium* **sp.**

Pl. 243

This bacterium induces small, irregular, globular root galls (nodules) on all lupines. These root nodules are 3 to 10 mm in diameter but may vary greatly in size depending on the host species. The surface of the galls appears smooth and usually light beige in color. This nitrogen-fixing bacterium has a symbiotic relationship with its host, extracting atmospheric nitrogen, storing it within the galls, and making it available to the host as well as other plants in the area (see the root nodule discussion in the section on bacteria).

Plate 244. The gall of *Dasineura lupini* on bush lupine.

Plate 245. The gall of *Dasineura lupinorum* on bush lupine.

LUPINE BUD GALL MIDGE   *Dasineura lupini*
Pl. 244

This midge induces large, globular, polythalamous, pubescent bud galls on yellow bush lupine *(Lupinus arboreus)* and blue bush lupine *(L. chamissonis)* in spring. Galls exposed to direct sun are usually red or purple, while galls completely shaded are green. Some galls reach 30 mm in length by 20 mm in diameter but are typically smaller. Galls are covered with a short, silver pubescence. They are composed of greatly expanded bud scales and often have dwarfed petioles and tiny leaflets emerging. Often the tips of bud scales recurve away from the main body of the gall. Galls occur singly or in small clusters, with several adjoining buds galled. The larval chamber is located centrally near the base of the gall. Melon orange larvae occupy galls in July. Larvae remain in the gall through fall and winter, pupating the following spring, timed with new bud development. Adults emerge in May.

LEAF-FOLDING MIDGE   *Dasineura lupinorum*
Pl. 245

This midge induces midrib fold galls on the leaflets of yellow bush lupine *(Lupinus arboreus)*, blue bush lupine *(L. chamissonis)*,

Plate 246. Integral stem gall of *Neolasioptera lupini* on bush lupine.

L. albifrons, L. albicaulis, and *L. variicolor* from Oregon to southern California. The galls are small, midleaflet swellings, or they incorporate the entire leaflet and often all leaflets on a petiole. The leaflets are completely folded together from edge to edge at the gall site. Galls measure from 3 to 15 mm in length by 2 to 3 mm in diameter. These galls are either a wine red or a gray green color. It has been suggested that those that turn red contain dead larvae, perhaps as a result of parasitism. In some cases, particularly during summer, as many as 80 percent of the leaflets on a host shrub are galled. As with many other *Dasineura* species, the larvae complete development inside the gall and remain there through pupation. This midge produces several generations per year. Fresh, larvae-containing galls can be found nearly year-round. Females lay their eggs on the unopened leaflets of growing buds. Newly hatched larvae crawl between the closed halves of the leaflets and begin feeding, stimulating the formation of the swollen galls.

LUPINE STEM GALL MIDGE   *Neolasioptera lupini*
Pl. 246

This midge induces integral, polythalamous galls on yellow bush lupine *(Lupinus arboreus)*, *L. variicolor,* and *L. densiflorus,* among others, from Washington to California. These galls can be evenly tapered and elliptical in shape on a straight stem, or they can bend the stem. Galls of this midge reach 30 mm in length by 10 mm in diameter. They are usually green like the stems. Leaves and stems can emerge from the sides of these galls. Most galls have a sparse pubescence, depending on the host species. As with the Leaf-folding Midge *(Dasineura lupinorum)*, larvae remain in the galls, and the adults emerge the following spring.

Manzanita Galls

California has 50 species of manzanitas (*Arctostaphylos* spp.), and additional species occur in other western states. Two known agents induce galls on this genus: one is a fungus, and the other is an aphid (Aphididae).

References: Westcott 1971; Sinclair et al. 1987; Johnson and Lyon 1991; Miller, D., 1998, 2000, 2004; Miller, W., and Sharkey 2000; Wool 2004.

MANZANITA LEAF FUNGUS   *Exobasidium vaccinii*
Pl. 247

This fungus induces large, bright red, swollen, convex galls on the dorsal surface of the leaves of many species of manzanita. The fungus can also induce witches' brooms. There is a corresponding depression on the underside of the leaf where the basidiospores are produced. Galls can encompass half to three-fourths of the affected leaf. These round to oval-shaped galls measure 15 mm long by 12 mm across. Larger specimens may occur. The basidiospores produced by this fungus are distributed by air currents and germinate on wet surfaces of host plants. New infections are caused when hyphae directly penetrate new leaves. The hyphae of this fungus spread throughout the entire host. As a result, galled leaves can appear all over the infected shrub. This fungus also occurs on cranberries and huckleberries (*Vaccinium* spp.), mountain heather (*Cassiope* spp.), rhododendron (*Rhododendron* spp.), Labrador tea *(Ledum groenlandicum)*, and dwarf blueberry *(V. caespitosum)*. This fungus is found across the United States and Canada.

Plate 247 (upper left). Pocket gall of *Exobasidium vaccinii* on manzanita.

Plate 248 (upper right). The galls of *Tamalia coweni* on manzanita.

Plate 249 (left). The flower galls of a wingless female *T. coweni* on manzanita.

MANZANITA LEAF GALL APHID   *Tamalia coweni*

Pls. 248, 249

This aphid induces two kinds of fold galls: one along the margins of leaves of many species of glabrous-leaved manzanitas, and the second on the inflorescences of manzanitas. These aphids will also induce midrib galls on certain host manzanita species. This aphid does not appear to occur on species with hairy leaves. Sometimes new leaves are completely galled by the fundatrices (stem mothers) of this aphid. The galls are succulent and pink red and measure 10 mm long by 5 mm across. Although most galls turn brown by fall, after the aphids have left, you can find fresh, occupied galls almost anytime of year. They often occur singly or in pairs on opposite sides of the leaf. Shrubs have been found with nearly every leaf galled by this aphid. The process begins in spring when the stem mother selects a row of cells, stinging the row in a pattern that causes the leaf to swell and fold over her.

A single gall can contain several stem mothers. Once enclosed, stem mothers produce offspring that fill the gall and mature into alates (see the discussion of aphid biology in the section on aphids and adelgids). These winged females seek developing inflorescences in June and July (which would blossom the following spring), producing a brood of wingless females. The wingless females resemble stem mothers and induce globular galls by converting flower buds into large, pink red, swollen bags. When several flower buds are galled, they look like red grapes. These alates mate and deposit the eggs that overwinter in the crevices of the bark at the base of the shrubs. By using next year's developing inflorescences as an alternate host site during early summer, these aphids apparently extend the gall-making season and their ability to complete their life cycle with the numerous stages involved. Also, these late-season galls tend to have lower rates of invasion and parasitism. The inquilinous aphid *Tamalia inquilina* is considered a commensal insect in that it appears not to significantly impact survival rates of the Manzanita Leaf Gall Aphid even though it does reduce its reproductive success. The inquiline, found only in California and Baja California, enters occupied as well as abandoned galls and reproduces successfully, which supports its consideration as a commensal. This aphid, *T. coweni,* is widespread in Colorado, Nevada, and the Pacific Coast states, as well as British Columbia.

Meadowsweet Galls

Meadowsweet (*Spiraea* spp.) is widely distributed across the United States. Even though several gall midges (Cecidomyiidae) are known to be gall inducers for these shrubs elsewhere, there is only one gall organism—a leaf-roller—on the two species of meadowsweet in California.

Reference: Gagné 1989.

MEADOWSWEET GALL MIDGE   *Dasineura salicifoliae*
Pl. 250
This midge induces swollen, polythalamous, midrib fold-roll galls on the leaves of *Spiraea densiflora* and *S. douglasii*. These long, swollen, fold galls are usually pink, rose, or red and quite

Plate 250. Roll galls of *Dasineura salicifoliae* on meadowsweet.

noticeable. Galls measure up to 25 mm long by 3 mm in diameter. Occasionally they roll the margins of the leaves. Galls protrude on the underside of the leaves in the midrib form. Each gall has a single large chamber within which several orange larvae feed and grow. Generally, these galls form during early summer and are fully developed by July. Larvae leave through a longitudinal slit and pupate in leaf litter the following spring. Another midge, *Parallelodiplosis* sp., is considered an inquiline in the galls of *Dasineura*. Little else is known about the biology of this midge. This species is also found in Oregon, Washington, and British Columbia.

Monkey Flower Galls

California has several species of monkey flowers (*Mimulus* spp.). Currently, only one species, the bush monkey flower (*M. aurantiacus*), also known as the sticky monkey flower, is known to host a gall organism. This diminutive shrub with its melon orange flowers dots the coastal and foothill landscapes throughout the state. The organism is a gall midge (Cecidomyiidae).

Reference: Gagné 1989.

MONKEY FLOWER STEM GALL MIDGE   *Neolasioptera*
Pl. 251   *diplaci*

This midge induces integral, elliptical, monothalamous stem galls. Galls occur midstem or at bud nodes. Node galls measure 7 mm long and wide and are abrupt swellings. Midstem galls measure 15 mm long by 7 mm diameter and are gradually tapered. The

Plate 251. Integral stem gall of *Neolasioptera diplaci* on bush monkey flower.

galls assume either the normal color of new growth stems, a reddish brick color, or are pale green. Node galls usually have leaf petioles or buds protruding from the galls. Adults emerge in fall. Little else is known about the biology of this midge. It has been found only in California.

Mormon Tea Galls

Seven species of Mormon tea (*Ephedra* spp.) occur in the arid regions of California. Many of these species also occur in other western states. This unusual plant spends most of the year leafless, using the chlorophyll in its stems for photosynthesis, until ample rains allow leaf production. Two known gall midges (Cecidomyiidae) gall the stems of Mormon tea.

Reference: Gagné 1989.

MORMON TEA STEM GALL MIDGE   *Lasioptera ephedrae*
Pl. 252
This midge induces dimpled, integral stem galls usually just above the nodes of desert tea *(Ephedra californica), E. nevadensis, E. tri-*

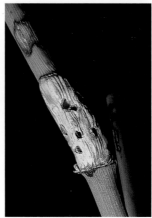

Plate 252. Integral stem gall of *Lasioptera ephedrae* on Mormon tea showing exit hole.

Plate 253. The gall of *Lasioptera ephedricola* showing several exit holes.

furca, and perhaps other western species. The monothalamous galls measure 10 to 15 mm long by 5 mm in diameter. Galls are stem green, taper gently, and show a dimple of thinner, beige tissue about midlength, where the larval chamber is located. The larvae feed in the pithy center of the stems. Pupation occurs in the galls, with the adults emerging the following spring through the thin wall of the dimple. There is one generation per year. This midge has been found from New Mexico to California.

EPHEDRA STEM GALL MIDGE   *Lasioptera ephedricola*
Pl. 253

This midge induces integral, wrinkled or furrowed stem galls on desert tea *(Ephedra californica), E. nevadensis,* and *E. trifurca.* These polythalamous galls measure 10 to 30 mm long by 4 mm in diameter. Galls are normally darker green than the rest of the stems. The larval tunnels are lined with a fungus (see the discussion of ambrosia galls in the section on gall midges). Pupation occurs inside the galls, with the adults emerging in February and March. This midge occurs from New Mexico to California and is common in the southern Mojave Desert.

Poison Oak Galls

With its delicate leaves of three and its brilliant hues of red, orange, and purple decorating hillsides in fall, poison oak *(Toxicodendron diversilobum)* may be the scourge of hikers and campers, but it is a boon to wildlife, including some gall organisms. Two galls occur on poison oak with regularity: an eriophyid mite gall on the leaves and a large, antler-shaped fasciation gall of the terminal buds with an unknown origin. A chocolate brown rust fungus, *Pileolaria effusa,* also attacks the petioles, causing significant swelling. The mite and fasciation galls are described here.

Reference: Keifer et al. 1982; Sinclair et al. 1987.

POISON OAK LEAF GALL MITE *Aculops toxicophagus*
Fig. 78

This mite induces red bead or pouch galls on the lower and upper surfaces of the leaves of both poison oak *(Toxicodendron diversilobum)* in the West and poison ivy *(T. radicans)* and poison sumac *(T. vernix)* in the East. Galls occur in a scattered, random manner across the surface of the leaflets. Sometimes individual galls (1 to 2 mm in diameter) combine with others to form large masses (5 mm in diameter). There is a corresponding depression and opening on the opposite side of the leaf from the pouch or bead gall, which allows escape of the mites later in the season. Although usually red in color, the pouch galls may also be green, pink, or brown. Galled leaves become distorted under heavy infestation. This mite occurs throughout the range of its host plants in California, Oregon, and Washington.

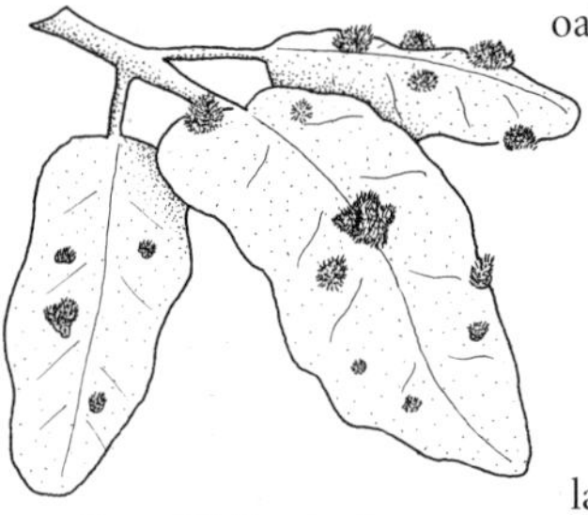

Figure 78. Bead galls of *Aculops toxicophagus* on poison oak.

Plate 254. Fasciation of the terminal buds of poison oak.

POISON OAK FASCIATION GALLS
Pl. 254

Undescribed

These galls are large, antler-shaped fasciations on terminal buds, and their broad, flattened shape stands out from normal branches. The galls measure up to 25 to 30 cm long by 15 cm wide and bend and twist in ways that tease the imagination. They often show small, distorted leaflets at the tips. During the second growing season, new shoots will elongate but will also produce fasciations at the terminal buds. This process seems to continue for three or four seasons before the branches die. Whatever the cause, entire patches of dead poison oak can be found after three or four seasons of fasciation. It appears that eriophyid mites (Eriophyidae) are not associated with these fasciations, leaving us to ponder other possible causes. This may well prove to be a "noninfectious fasciation" specific to isolated clones of poison oak.

Plum Galls

California has seven species of native shrubs and small trees in the genus *Prunus,* including choke-cherry *(P. virginiana),* holly-leafed cherry *(P. ilicifolia),* and Sierra plum *(P. subcordata).* Some species share the same gall organisms, while other species have their own unique gall inducer. The galls normally found on wild plums in the West include eriophyid mites (Eriophyidae) and ascomycete fungi in the genus *Taphrina.* A different ascomycete fungus, black knot gall fungus *(Apiosporina morbosa),* distorts stems of several species of *Prunus.* An unknown gall moth induces large, midrib, leaf-fold galls that show mostly on the underside of choke-cherry leaves in the central Sierra.

References: Mix 1949; Westcott 1971; Russo 1979; Keifer et al. 1982; Sinclair et al. 1987.

WITCHES' BROOM FUNGUS   *Taphrina confusa*
Pl. 255

This fungus induces massive witches' brooms on several species of *Prunus,* but most particularly western choke-cherry *(Prunus virginiana* var. *demissa)* and Sierra plum *(P. subcordata)* here in the West. These brooms are easily seen while driving along a road, as the clusters of compact branches exceed 25 cm (10 in.) across. Each broom has dozens of branches with scattered, stunted leaves. Even though this fungus has been reported across the United States, it appears to have localized concentrations. The area around Mount Shasta, for example, has a significant concentration of broomed *Prunus.* The taxonomy of this fungus changes among authors. The name used by Westcott (1971) is adopted here.

PLUM LEAF-CURL FUNGUS   *Taphrina flectans*
Pl. 256

This fungus induces leaf-curl galls on new spring leaves at the tip of branches and witches' brooms on bitter cherry *(Prunus emarginata)* and holly-leafed cherry *(P. ilicifolia).* One or more leaves may be galled in a cluster, severely distorting and discoloring them. Galled leaves are normally swollen, yellow green, and turned under at the edges. These leaves usually turn brown be-

Plate 255 (upper left). Witches' broom caused by *Taphrina confusa* on choke-cherry.

Plate 256 (upper right). Leaf curl caused by *Taphrina flectans* on holly-leafed cherry.

Plate 257 (left). Bladder plums caused by *Taphrina prunisubcordatae* on Sierra plum.

fore the end of the normal growing season, shortly after the asci have released their spores. This fungus goes through a saprophytic stage in which spores bud on plant surfaces. Some of these spores survive winter on bud scales and germinate in spring as new leaves unfold. The taxonomy of *Taphrina* on *Prunus* spp. needs clarification.

PLUM POCKET GALL FUNGUS

Taphrina prunisubcordatae

Pl. 257

This fungus induces "plum pockets," also called "bladder plums," on Sierra plum *(Prunus subcordata)*. Plum pockets are infected fruits that expand and elongate into large, soft, hollow, furrowed, light green bags that are round to potato shaped. Under the influence of this fungus, normal fruit measuring 10 to 12 mm across swell into fungus galls, measuring 45 mm long by 20 mm wide.

In spring, these galls might be mistaken for fruit, as they hang in clusters like other fruit. With age, the galls turn beige and dry. The surface of the galls is somewhat mealy-powdery with a waxy bloom. While the fungus appears to abort attacked fruit, it does not appear to eradicate the entire fruit production on a single affected shrub. This fungus occurs from the Rocky Mountains west, throughout the Pacific states. The fungus *T. pruni* produces enlarged fruits on *P. americana* and *P. domestica* in the West.

BLACK KNOT GALL FUNGUS *Apiosporina morbosa*
Fig. 79 **(syn. *Dibotryon morbosum*)**

This fungus induces massive eruptions or knot galls on the stems of several species in the genus *Prunus*, both native and ornamental. These integral stem galls can be found anytime of the year in various conditions. They range in size from 6 to 16 cm long by 3 cm in diameter. The elongated, bent, sometimes twisted black galls make their first appearance on branches in fall after the initial infection the previous spring. Full development is hindered by winter, until the following spring when normal growth resumes. The fungus secretes indoleacetic acid and other growth-regulating compounds that stimulate the development of the knot gall. In early summer of the second season of development, galls are olive green and covered with conidiospores. Ultimately, ascospores are produced as the fungus enters another stage in its life cycle. As the galls mature and release all of their spores, the galls become pitted, cracked, and knobby.

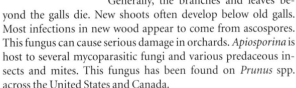

Figure 79. Black knot gall of *Apiosporina morbosa* on Sierra plum.

Generally, the branches and leaves beyond the galls die. New shoots often develop below old galls. Most infections in new wood appear to come from ascospores. This fungus can cause serious damage in orchards. *Apiosporina* is host to several mycoparasitic fungi and various predaceous insects and mites. This fungus has been found on *Prunus* spp. across the United States and Canada.

Plate 258.
Nail galls of
*Phytoptus
emarginatae*
on Sierra
plum.

BITTER CHERRY GALL MITE   *Phytoptus emarginatae*
Pl. 258

This mite induces clublike pouch galls on leaves of several species of *Prunus*, especially bitter cherry *(P. emarginata)*. These blunt-tipped, minutely pubescent spring galls measure up to 5 mm high by 1 mm in diameter. Galls occur on both surfaces of the leaves, with several dozen galls per leaf being common. A microscopic opening leads into the gall chamber on the opposite side of the leaf from the erect club. When fresh, the galls are beige, green, or red depending on their exposure and age. While most galls stand erect, some bend over or twist. There is a single generation of these mites each year. Females retire to crevices in terminal branches by late summer. After contacting males, females carry sperm with them through winter. Females induce the galls the following spring that contain both males and females. This mite is widespread across North America and in Europe. Damage to leaves and shrubs appears to be negligible, although some browning of leaf tissues may occur under heavy infestation.

Rabbitbrush Galls

Rabbitbrush (*Chrysothamnus* spp.) often decorates arid land-scapes with its flowers at a time of year when few other plants are in bloom. It is one of the keystone plants in the eastern Sierra, Great Basin country, and areas of the Mojave Desert, as well as other arid regions in the West. California has over 10 species of rabbitbrush. Rabbitbrush, like creosote bush *(Larrea tridentata)* and saltbush (*Atriplex* spp.), is one of those shrubs that supports an unusually large number of gall-inducing insects. This important small shrub hosts several species of gall midges (Cecidomyiidae), tephritid fruit flies (Tephriditae), and a moth (Gelechiidae). Several new species have yet to be identified. Many of these gall insects initiate gall formation during fall and winter when most other galls have stopped growing. Unlike the galls of cynipid wasps (Cynipidae), the galls induced by tephritid fruit flies can vary in shape from one region to another and from one species of rabbitbrush to another. Together, rabbitbrush and its interesting gall insects and their associates represent a unique relationship in the world of plant galls.

References: Pritchard 1953; Foote and Blanc 1963; Tauber and Tauber 1968; Wangberg 1980, 1981a,b; Gagné 1989; Goeden et al. 1995; Goeden and Teerink 1996a,b,c, 1997a,b,c; Headrick and Goeden 1997; Headrick et al. 1997.

BENT GALL MOTH *Gnorimoschema octomaculellum*
Pl. 259

This moth, in the family Gelechiidae, induces pointed, monothalamous galls out of the terminal leaves on new shoots in spring on rubber rabbitbrush *(Chrysothamnus nauseosus)*. The galls are typically bent at a 45-degree angle to the supporting stem. Galls are composed of swollen, succulent, distorted terminal leaves that are tightly bound together like the staves of a barrel. There is no growth beyond the galls. In some areas, caterpillars emerge by late spring through exit holes in the sides of the galls. These caterpillars form sediment-covered cocoons in the substrate below the host shrub that measure 10 to 15 mm long and 2 to 3 mm wide. Rearing experiments indoors showed adults emerging in late May. Under natural conditions, the timing for these events may be much later. Old, brown galls remain for several months on dead stems.

COTTON GALL TEPHRITID *Aciurina bigeloviae*
Pl. 260

This fly induces round-globular, woolly, white, monothalamous bud galls on rubber rabbitbrush *(Chrysothamnus nauseosus)*. These compact clusters of white hair usually measure 12 to 15 mm

Plate 259.
Terminal leaf gall of
*Gnorimoschema
octomaculellum*
on rabbitbrush.

Plate 260.
Bud galls of
Aciurina bigeloviae
on rabbitbrush.

in diameter. Galls occur singly or in pairs at axillary buds. Galls appear as puffs of cotton attached to the stems and are usually found on the current year's growth. Old galls persist on the branches for a year or more. Fresh galls in the eastern Sierra have been found in September and October with larvae. The adults emerge in spring. Adult behavior is probably like that of other members of the genus. The galls of this species are used by gall midge *Rhopalomyia bigeloviae* to create its own "endogalls," as it does with the Bubble Gall Tephritid *(A. trixa)*, described on rabbitbrush later.

Plate 261. Bud gall of *Aciurina ferruginea* on rabbitbrush.

Plate 262. Bud galls of *Aciurina idahoensis* on rabbitbrush.

MEDUSA GALL TEPHRITID   *Aciurina ferruginea*

Pl. 261

This fly induces large, leafy, monothalamous bud galls on the stems of yellow rabbitbrush (*Chrysothamnus viscidiflorus*). The galls are characterized by numerous, long, thin, leafy bracts, which appear to develop in fascicles from many small buds. Galled leaves range in size from 3 to 18 mm in length. When fresh, galls are green, turning brown after the larvae stop feeding. Fresh galls have been found in September in the eastern Sierra. Sometimes galls close together combine into one long, leafy mass on the side of a stem, reaching 60 mm in length. Larvae cut a tunnel to the outer skin of the gall, leaving a thin layer of skin to cover the tunnel beneath. Pupation occurs within the galls. Adults break through the thin skin to emerge and then push aside the filamentous leaves to gain access to the outside in July and August. Adults mature within 10 days or so, and eggs are deposited shortly thereafter. A single egg is deposited on each axillary bud. Gall formation begins after the eggs hatch and larvae commence feeding. Adults live up to 105 days and on average about 44 days. There is one generation per year. This tephritid overwinters in the first-instar larval stage. This species has been found in all western states on its host. Two species of wasps in the genus *Halticoptera* are the primary parasites of this tephritid.

LEAFY CONE GALL TEPHRITID   *Aciurina idahoensis*
Pl. 262

This fly induces round-globular, monothalamous, pubescent, leafy bud galls on yellow rabbitbrush *(Chrysothamnus viscidiflorus)*. Galls have numerous relatively thin, leafy bracts (bud scales) that stick out and sometimes recurve from the main gall body. The galls are usually green with white hairs over the leafy bracts. Galls occur singly or in rows bunched together along the length of the stem. Individual galls measure up to 20 mm in length by 8 mm in diameter. The exterior leafy bracts measure 10 mm long by 1 mm in width (these bracts can be much longer, based on local variation). The central larval chamber is vertical and measures up to 8 mm long by 2 mm wide. Pupation occurs in the galls, with the adults emerging in the spring through the gall apex by pushing the apical tissues aside. Adults live up to 22 days. Egg deposition occurs shortly after emergence and mating. There is one generation per year. Several hymenopteran parasites are associated with this fruit fly, including *Eurytoma chrysothamni*, *Platygaster* sp., *Halticoptera* sp., and a *Torymus* sp.

BEAKED GALL TEPHRITID   *Aciurina michaeli*
Fig. 80

This fly induces monothalamous, pear-shaped, smooth bud galls on the current year's growth of rubber rabbitbrush *(Chrysothamnus nauseosus)* and yellow rabbitbrush *(C. viscidiflorus)*. Galls are composed of numerous undeveloped bud leaves that are incorporated into the main body of gall tissue. Usually, a few leafy bracts on the outside come to a point at the tip of the galls. A subtle pubescence may occur over the surface of the exterior bud leaves. Galls measure 10 to 12 mm long by 4 to 6 mm in diameter and are leaf green when fresh. Toward the end of larval growth, the larvae concentrate their feeding in the apical end of the galls, thinning the tissues. The adults later cut through this area of thin tissue to exit the gall. Larvae pupate in the galls, and adults emerge in July and August. There is one generation per year. This species has been reported from nearly all western states.

Figure 80. Bud gall of *Aciurina michaeli* on rabbitbrush.

Plate 263 (left). Normal bud galls of *Aciurina trixa* on rabbitbrush.

Plate 264 (right). Gall of *A. trixa* showing an endogall of the midge *Rhopalomyia bigeloviae* protruding on rabbitbrush.

BUBBLE GALL TEPHRITID   *Aciurina trixa*

Pl. 263, 264

This fly induces small, round to globular, monothalamous stem galls on rubber rabbitbrush *(Chrysothamnus nauseosus)*. While this fruit fly may appear on other species of rabbitbrush, it seems to prefer this species for galling. The galls of the Bubble Gall Tephritid have been reported by Wangberg (1980) to vary greatly in morphology based on host and locality, but the galls found most often have the following characteristics. In midwinter in the Tehachapi Mountains, galls are fresh, lime green, round to oval, smooth, and glossy. They measure 5 mm in length by 3 to 4 mm in diameter and appear on the lower portions of branches. In some regions, such as Idaho and the northern Mojave in March, these galls can be over 10 mm in diameter. Most galls are glabrous and sticky. While many galls are smooth, some others have small, blunt-tipped protruberances, which are associated with gall midges (Cecidomyiidae) (see below). As the galls age, they turn pale green, yellow, and ultimately brown. Galls develop from axillary buds beginning in August at the initiation of feeding larvae. By November, most galls are full sized. These galls form at a time when the host plants are generally not actively growing. Galls aborted after the death of the larvae will often have shoots growing through them. Many host plants have more than 75 percent of their stems galled by this fly. Larvae create an exit tunnel before pupation. Adults emerge in spring and, like other tephritids, use wing displays in courtship. Single eggs are placed near the base of leaves. Larvae are active in galls by August, and they overwinter in the third instar. Pupae are found from March to June. This fruit fly may be found throughout the range of its host plants in several western states. Sixteen species of insects have been found associated with the galls of this fruit fly. One of the associates represents an incredibly rare case where one insect galls the galls of another species. The knobby protruberances mentioned earlier are actually the endogalls of a gall midge, *Rhopalomyia bigeloviae*. The galls of this midge protrude 1 to 2 mm, look like knobs or fingerlike projections, and take on the characters of the host gall, that is, glabrous and sticky. Growth and development of the midge and the tephritid larvae proceed without any apparent impact of one upon the other. Both species pupate and emerge about the same time. While several tephritids have been known as inquilines of gall midge galls, this is a rare situation and one of the most intimate relationships between gall-inducing insects known. Apparently, this midge also galls the fuzzy galls of the Cotton Gall Tephritid *(A. bigeloviae)*.

LEAFY BUD GALL TEPHRITID *Procecidochares* sp. A
Pl. 265, Fig. 81

This fly induces leafy, pubescent, monothalamous bud galls on *Chrysothamnus humilis* and yellow rabbitbrush *(C. viscidiflorus)*. Galls of this species are normally associated with the *P. minuta* group. They occur singly or in rows along the length of the stems. Broad-based leafy bracts that measure 10 mm long by 8 mm wide at the base distinguish this species from other related forms. Galls measure up to 31 mm long, the average being around 17 mm long by 15 mm wide. They are composed of overlapping bud leaves that are basally attached to the main gall chamber. Some galls look like small artichokes and are often sticky.

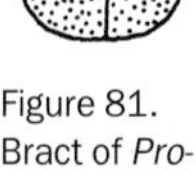

Figure 81. Bract of *Procecidochares* sp. A bud gall.

These galls are green with a sparse pubescence. Once the larvae stop feeding, the galls turn brown. Adults emerge during the morning hours through the thin apical tissue of the chambers and by pushing through leafy bracts at the apex. Courtship in-

Plate 265. Bud galls of *Proceci-dochares* sp. A on rabbitbrush.

volves complex wing displays, posturing, and orientation combined with movement of mouthparts. Females oviposit directly into axillary buds, with one egg per bud in spring. There is one generation per year. Nine insect associates have been recorded for this fruit fly including two beetles, two moths, a gall midge, and four wasps.

HAIRY BUD GALL TEPHRITID   *Procecidochares* sp. B
Pl. 266, Fig. 82

This fly induces galls similar to those of the Leafy Bud Gall Tephritid (*Procecidochares* sp. A), except the galls of this species are gray green and densely covered with white pubescence, have shorter bracts, and usually occur only on terminal buds of rubber rabbitbrush *(Chrysothamnus nauseosus)*. This species has also been associated with the *P. minuta* group. These galls measure up to 20 mm long by 10 mm wide. Most, however, are 12.5 mm long by 5.6 mm wide. Bracts measure

Figure 82. Bract of *Procecidochares* sp. B bud gall.

12 mm long by 5 mm wide at the base and have a short apex, not long and pointed as in the Leafy Bud Gall Tephritid. In addition, a distinctive, fibrous mesh of hairs connects the leafy bracts along the margins and at the tips. Gall development begins in spring in concert with new leaves. As the galls grow, the immature leaves composing the galled buds fuse together at the base, forming the rigid larval chamber. Prior to adult emergence and after larvae have stopped feeding, galls begin to desiccate and turn brown. The drying process causes the tip of the leafy bracts to separate, which allows easy escape for the emerging adults later. By the end of summer, the vacated galls have withered and most have

Plate 266. Bud galls of *Procecido-chares* sp. B on rabbitbrush.

dropped off the host plant. Adults are usually found in June and July and mate soon after emergence, following courtship behaviors similar to the Leafy Bud Gall Tephritid. Females usually oviposit in the buds nearest to the ones they emerged from but rarely deposit eggs in more than four buds before interruption by ants and spiders. Eggs appear to overwinter, and hatching occurs in spring. Larvae are within newly formed gall tissue by late April. Each gall contains from one to three larvae with as many as six in a gall. There is one generation per year. Eight species of insect associates have been found in these galls, as inquilines, parasites, or hyperparasites.

ELLIPTICAL STEM GALL TEPHRITID *Valentibulla californica*
Pl. 267

This fly induces integral, barely swollen, polythalamous stem galls on rubber rabbitbrush *(Chrysothamnus nauseosus)*. These galls may taper gently or abruptly and may be symmetrical or lopsided. Galls measure 30 mm long by 6 mm in diameter and are only slightly larger than the stem. They are colored like the stems. Lopsided galls may have random bulges, which correspond to feeding larvae. Each larva occupies its own chamber. Before pupation, each larva excavates a tunnel to the surface, leaving the epidermis as a shield from the outside. Most of the year is spent in larval stages. Adults emerge in June in some areas. These flies have a single generation per year.

STEM GALL MIDGE *Rhopalomyia chrysothamni*
Pl. 268

This midge induces small, conical-tubular galls that project out at right angles to the stems of rubber rabbitbrush *(Chrysotham-*

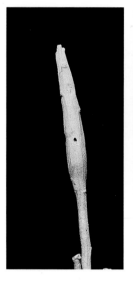

Plate 267 (left). Integral stem gall of *Valentibulla californica* on rabbitbrush.

Plate 268 (right). Tubular galls of *Rhopalomyia chrysothamni* on rabbitbrush.

nus nauseosus). Tubular clusters of white hairs protrude from the galls. The galls occur singly or in groups. They are 3 mm long by 2 mm in diameter. There is generally some swelling at the base where the galls emerge from the stems. Often several galls protrude in different directions from the same section of the galled stem. In other cases, several galls emerge from the same side of the stem, causing the stem to bend in the opposite direction. Old galls remain on the dead stems. These midges have been found from Utah to southern California.

JELLYBEAN GALL MIDGE   *Rhopalomyia glutinosa*
Pl. 269

This midge induces globular, shiny, black or lime green, monothalamous stem galls near the base of new spring growth and just above the connection with last year's branches of rubber rabbitbrush *(Chrysothamnus nauseosus).* These jellylike but hard, flat-topped, glutinous blobs measure 3 mm high by 4 mm wide. Some galls have small protuberances over the top, with green coloration around the base. Several galls may occur in close proxim-

Plate 269. The bud gall of *Rhopalomyia glutinosa* in its fresh, green form on rabbitbrush.

Plate 270. The bud gall of *Rhopalomyia utahensis* on rabbitbrush.

ity to each other but apparently do not coalesce, as do other species. Galls found in late November were glossy black, while specimens seen in March were lime green. This midge has been found from Utah west into the Tehachapi Mountains and the Mojave Desert.

COTTON GALL MIDGE   *Rhopalomyia utahensis*
Pl. 270

This midge induces white, cottony-pubescent, oval, monothalamous galls along the stems of rubber rabbitbrush *(Chrysothamnus nauseosus)*. Galls develop on axillary buds in spring. These galls measure up to 26 mm long by 12 mm in width. Galls are normally covered with numerous recurved leaflets, which are the tips of the galled bud scales or leaflets. When fresh, galls are usually green with a white tomentum between the recurved leaflets. Sometimes the galls have a purplish hue. With age, the galls turn brown. Pupation occurs in the galls, with the adults emerging the following April or May depending on location. Adults mate soon after emergence and live for only one or two days. Either the eggs or first-instar larvae overwinter in diapause until the following spring. This midge has been found in Idaho, Utah, and California but could occur elsewhere within its host's range.

Plate 271. Fresh bud gall of *Rhopalomyia* sp. on rabbitbrush showing the distinctive hooks.

Plate 272. Old gall of *Rhopalomyia* sp. showing a pupal case at the surface.

SPINY CONE GALL MIDGE   *Rhopalomyia* sp.

Pls. 271, 272, Fig. 83

This midge induces glabrous, purple rose, cone-shaped stem galls on rubber rabbitbrush *(Chrysothamnus nauseosus)* and per-haps other species as well. These wine-colored galls can be either monothalamous or polythalamous. The flat top and the recurved, green, pointed bracts that are sparsely arranged along the sides distinguish them. While these galls may superficially resemble those of the Cotton Gall Midge *(R. utahensis)*, they lack the white hairs characteristic of those galls. The galls of this species begin development in early spring and are fully formed by mid-March. Galls measure 7 mm high by 5 mm in diameter. Coalesced galls (see fig. 83) measure 15 mm in diameter by 10 mm high. Pupae develop in a hole in the center of the flat-topped surface. By July, these extremely hard galls are brown. They may remain on the stems for a year or more.

Figure 83. Cross section of two *Rhopalomyia* sp. galls showing the position of the larval chambers.

Plate 273. Bead galls of *Aceria boycei* on ragweed.

Ragweed Galls

Several species of ragweed (*Ambrosia* spp.), also called bur-sage, are found in the West. They occur in a variety of habitats from coastal dunes to desert and are famous for allergenic pollen. The gall organisms associated with these shrubs include eriophyid mites (Eriophyidae), tephritid fruit flies (Tephritidae), and gall midges (Cecidomyiidae).

References: Keifer 1952; Gagné 1975, 1989; Silverman and Goeden 1980; Goeden and Teerink 1997a.

RAGWEED GALL MITE   *Aceria boycei*
Pl. 273
This mite induces round-globular, convoluted bead galls on the dorsal surface of the terminal leaves of western ragweed (*Ambrosia psilostachya*). The exit holes are located on the undersurface of the leaves corresponding to the bead galls above. Galls occur singly but will coalesce to form larger, warty clusters on the leaves. Sometimes, leaves are severely distorted. Galls measure 2 to 4 mm in diameter and are usually yellow green. This mite is found primarily on hosts in southern California.

Plate 274. Bud gall of *Proceci-dochares kristineae* on ragweed.

Plate 275. The floral bud gall of *Procecidochares lisae* on woolly bur-sage.

BUD GALL TEPHRITID   *Procecidochares kristineae*
Pl. 274

This fly induces leafy, pubescent, monothalamous bud galls on burro-weed *(Ambrosia dumosa)*. Galls measure 10 mm high by 7 mm in diameter and are gray green, often with hints of pink or rose red near the base. Galls usually develop timed with vegetative growth induced by substantive seasonal rains during winter and summer. Gall development, and therefore larval growth, may be arrested should adequate rains not occur during the season following initiation of gall formation. One study showed that first-instar larvae entered a "quiescent period" for over five months (April to September) during the absence of rainfall but resumed growth within 10 days following a major rainstorm. Adults usually mate and lay eggs within three days after emergence. Normally, there are two generations per year, but it can be one during arid periods. Nine species of parasitic wasps have been reared from these galls. Spiders and birds eat adults, and rodents chew through galls to get at the larvae. A similar tephritid gall induced by the Woolly Flower Gall Tephritid *(P. lisae)* occurs on woolly bur-sage *(A. eriocentra)* in southern California.

WOOLLY FLOWER GALL TEPHRITID   *Procecidochares lisae*
Pl. 275

This fly induces round, white-pubescent, monothalamous galls on woolly bur-sage *(Ambrosia eriocentra)* in the Mojave Desert

Plate 276. Bud gall of *Asphondylia ambrosiae* on annual bur-sage along the coast.

and perhaps elsewhere in the range of this host. Galls form just below the pale yellow flower heads, with some leafy bracts emerging from the sides of the galls or from normal axillary buds. These galls measure 12 to 14 mm high by 10 mm in diameter. The larval chambers are rather large, occupying most of the thin-walled, hard galls. In April, mature, white larvae measure 4 mm long. By the end of April black pupae have formed, with adults emerging soon after. Little else is known about this species.

AMBROSIA BUD GALL MIDGE *Asphondylia ambrosiae*
Pl. 276

This midge induces terminal, polythalamous, densely pubescent bud galls on annual bur-sage *(Ambrosia acanthicarpa)*, common ragweed *(A. artemisiifolia)*, and burro-weed *(A. dumosa)*. These rather large, globose, integral swellings measure 20 mm long by 17 mm wide. Galls are gray green, similar to the normal stems and leaves of the host. They appear to be more densely pubescent than normal plant parts. Galls are composed of several overlapping bud leaves with leaf petioles emerging from the tip of the gall's leafy bracts. The larval chamber wall is lined with a white fungus, as with many *Asphondylia* galls (see the discussion of ambrosia galls in the section on gall midges). Several pale orange larvae are found inside galls in mid-May. Pupation occurs within the galls. Adults emerge the following spring, mate, and oviposit. Galls are well formed by mid-May.

Plate 277. Tubular galls of *Contarinia partheniicola* on bur-sage showing the flared cup.

Plate 278. Bud gall of *Rhopalomyia* sp. on woolly bur-sage.

TRUMPET GALL MIDGE
Contarinia partheniicola
Pl. 277

This midge induces small, green, trumpet-shaped, monothalamous galls at the base of leaves and in bud axils of several species of ragweed, especially burro-weed *(Ambrosia dumosa)*. Galls are pubescent, particularly around the rim of the bowl or funnel. Galls measure 5 mm high by 3 to 4 mm wide at the top. These galls are often concealed among crowded new leaves, but the open funnel shape usually gives them away. Adults have been found in spring and fall. This gall midge has been recorded in Florida, Texas, and California.

WOOLLY BUD GALL MIDGE
Rhopamlomyia sp.
Pl. 278

This midge induces white, woolly lateral, polythalamous bud galls on woolly bur-sage *(Ambrosia eriocentra)* in the Mojave Desert. These spring galls bear stiff brown spines or modified leaf bracts that protrude through the tufts of white hair. Galls measure 15 mm high by 20 mm in diameter. Galled shrubs can have several dozen galls, with all of the buds of some branches affected. This new species was discovered in 2004 during the fieldwork for this guide. Little is known about its biology.

Rose Galls

Several species of shrubby roses (*Rosa* spp.) are found in the West along with numerous hybrids and ornamentals that have established themselves in the wild. The gall fauna on roses in the Pacific states is rich and in need of additional studies. For the most part, the gall organisms associated with roses are cynipid wasps (Cynipidae) in the genus *Diplolepis*, whose larvae are responsible for gall formation, as with other cynipids. These wasps have figured prominently in studies of basic gall biology and nutrient accumulation (see the discussion in the section "Galls as Nutrient Sinks"). I would expect that there are several more species of *Diplolepis* wasps, and their galls, in California and the West than described here. An isolated patch of roses I found among pines in the eastern Sierra, for example, supported seven species of *Diplolepis* galls.

References: Weld 1957; Shorthouse 1973a,b, 2001; Dailey and Campbell 1973; Shorthouse and Rohfritsch 1992; Shorthouse 1993; Brooks and Shorthouse 1998a,b; Shorthouse and Brooks 1998.

MOSSY ROSE GALL WASP   *Diplolepis bassetti*
Pl. 279, Fig. 84

This wasp induces large, round, bristly, detachable, polythalamous galls on Nootka rose *(Rosa nutkana)*, pine rose *(R. pinetorum)*, and interior rose *(R. woodsii)*, among other species and hybrids. These bud galls look like balls of moss or hair. Galls are

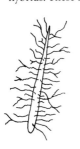

covered with hundreds of finely serrated bracts that distinguish this species from the galls of the Pincushion Gall Wasp *(D. rosae)* (compare fig. 84 and fig. 87). Galls measure to 50 mm in diameter and have a hardened core with bristly, mosslike hairs up to 20 mm long. In June, larval chambers are 3 mm long and are clustered centrally. Larvae are plump and white. The soft, yet bushy clusters of bristly bracts are brick red, red, yellow, or yellow green. Adults emerge in February and March. Mating and oviposition occur soon thereafter, depending on elevation. Well-

Figure 84. Bract of *D. bassetti* gall showing the long marginal hairs.

Plate 279. The mossy galls of *Diplolepis bassetti* on wild rose.

developed galls have been found at 1,200 m [3,936 ft] in the northern Sierra in May. This species has been found in several locations in the central and northern Sierra of California, Oregon, and Washington.

SPINY BUD GALL WASP *Diplolepis bicolor*
Pl. 280

This wasp induces round, spiny, monothalamous bud galls along the stems of several species of wild rose. The spines on this gall are stiff and scattered across the surface in an open arrangement mostly concentrated on the upper half of each gall. Galls measure 10 to 12 mm across with spines. Their color ranges from reddish to yellow green, depending upon exposure to sun. When old in late summer and fall, galls turn beige or dark brown. Galls may persist on the stems for more than one season. These galls look similar to those produced by the Spiny Leaf Gall Wasp *(D. polita)* on the upper surface of rose leaflets. Adults of the Spiny Bud Gall Wasp emerge in spring to oviposit in buds.

LEAFY BRACT GALL WASP *Diplolepis californica*
Pl. 281

This wasp induces green, leafy, polythalamous bud galls in spring on several species of wild rose. The galls of this wasp are distinguished from all others by the flat, leafy lobes that emanate from the main gall body and look like aborted leaflets. The lobes are up to 12 mm long by 2 mm wide at the tip. The entire gall mass

Plate 280 (upper left). Bud galls of *Diplolepis bicolor* on wild rose.

Plate 281 (upper right). Leafy bud gall of *Diplolepis californica* on wild rose.

Plate 282 (left). Inquiline modified gall of *Diplolepis nodulosa* on wild rose.

measures 25 to 38 mm in diameter. When green and fresh, the galls are soft and nonbrittle. Each gall can contain up to eight larval chambers surrounded by the white flesh of the gall. Adults emerge February to April, mate, and oviposit into spring buds.

ROSE STEM GALL WASP   *Diplolepis nodulosa*
Pl. 282, Fig. 85

This wasp induces tapered, elliptical, monothalamous, slightly swollen stem galls on wild roses and hybrids. These galls are usually located at the lower end of the affected stem, with shoot elongation beyond the galls. These galls measure 12 to 15 mm long by 3 to 4 mm in diameter. They often have stunted leaflets scattered

Figure 85. Normal gall of *Diplolepis nodulosa*.

around the periphery. While these small galls are often difficult to locate, galls modified by the common inquiline cynipid *Periclistus pirata* are easily seen because they are much larger, more abrupt swellings (pl. 282). Galls altered by *P. pirata* are round to ovoid, smooth, and glabrous and look like potatoes with dimples and creases. These galls measure up to 20 mm in diameter. Nearly half of the Rose Stem Gall Wasp galls examined in one study had been modified by the inquiline *P. pirata*, which has multiple larval chambers unlike the galls of its hosts. The larvae of the Rose Stem Gall Wasp are killed when *P. pirata* deposits its eggs. Both insects overwinter as prepupae, emerge in spring, and have one generation per year. Adult inquilines usually emerge two weeks after the emergence of the Rose Stem Gall Wasp, egg deposition, and the beginning of gall development.

SPINY LEAF GALL WASP   *Diplolepis polita*
Pl. 283, Fig. 86

This wasp induces round, spiny, pea-sized monothalamous galls on the dorsal surface of California rose *(Rosa californica)*, wood rose *(R. gymnocarpa)*, and interior rose *(R. woodsii)* leaves. These detachable, spring galls measure up to 5 mm in diameter, with several scattered spines emanating from the sides and top. Some galls coalesce, forming paired oblong galls. Galls occur singly or in large numbers per leaflet. Sometimes, all of the leaflets on a branch are galled, leaving little evidence of the leaflets themselves and making them look like stem galls. Galls are wine red, green, and yellow green. When fresh, the gall spines are erect and flexible. With age and by late summer, the spines are stiff and brittle, and the galls turn brown. Normal galls have fairly large larval chambers with thin outer wall tissue surrounding the central chamber. Parasitized galls usually have either several small larval chambers or a single chamber with an irregular inner wall. Adult emergence occurs in spring over a 2 to 6 week period, sometimes longer if the weather is cool. Reproduction is occasionally parthenogenetic for this and other species of *Diplolepis* as a result of *Wolbachia*

Plate 283. Spiny galls of *Diplolepis polita* on wild rose.

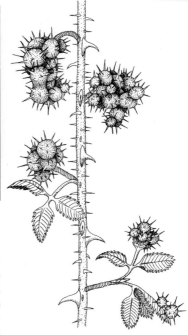

Figure 86. Clusters of *Diplolepis polita* galls showing full involvement of leaflets.

infection (a bacterialike organism), which causes the wasps to lay only female eggs. Research has shown that infected wasps can be cured through antibiotics; thereafter they lay both male and female eggs. This species has been found all over California but may occur in Oregon and Washington. Five species of parasites and inquilines are associated with the galls of this species, including the inquiline *Periclistus pirata* mentioned earlier. The others are *Eurytoma longavena, Glyphomerus stigma, Torymus bedeguaris,* and *Habrocytus* sp.

PINCUSHION GALL WASP *Diplolepis rosae*
Pl. 284, Fig. 87

This wasp induces round to ovoid, bristly, monothalamous galls on the dorsal surface of the leaflets of several European rose species and hybrids. These gall wasps have been accidentally introduced from Europe, where their galls are commonly known as the "rose bedeguar" or "robin's pincushion" galls. In Europe, this

Plate 284. Mossy leaf gall of *Diplolepis rosae* on wild rose.

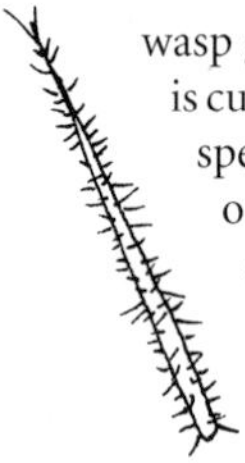

Figure 87. Bract of *D. rosae* gall showing the shortness of the marginal hairs.

wasp galls over 14 rose species, while in North America it is currently known from only a few naturalized European species, including *Rosa canina* and *R. eglanteria*. It could occur on other species and hybrids. These galls cover a portion or all of a leaflet (the similar galls of the Mossy Rose Gall Wasp [*D. bassetti*] are not on leaves but form on stems). The "hairs" or bristles are 8 mm long and have barely detectable, short side bristles, in contrast to those of the Mossy Rose Gall Wasp (compare fig. 84 and fig. 87). These galls are actually composed of several small (4 mm long), oblong, bristly, monothalamous galls that develop close together, creating the appearance of a single large gall (25 to 30 mm long). When fresh, the bristles are yellow green, beige, soft, and flexible. By late summer the galls dry and turn brown, and the hairs become brittle. Galls of this species appear to be more abundant on stressed plants than on those growing under optimal conditions. These galls have been found in the eastern Sierra and are common in Ontario, Canada, and presumably may occur elsewhere in the range of the hosts. Adults emerge in spring and deposit several eggs in each leaflet.

Plate 285. Blister-like integral leaf galls of *Diplolepis rosaefolii* on wild rose.

ROSE BLISTER GALL WASP *Diplolepis rosaefolii*
Pl. 285   **(syn. *D. lens*)**

This wasp induces round-lenticular, integral, swollen, blisterlike, monothalamous galls that show on both sides of the leaves of several species of rose. Galls are usually 2 to 3 mm in diameter, but some can reach 5 mm. Galls are lenticular when viewed from the side. Color ranges from green to yellow and dark red. Each leaflet can have one to several galls. Sometimes, several galls coalesce to completely distort the host leaflet. These gall inducers overwinter in a prepupae state. Pupation occurs from April to early June. The adults are known to be the smallest in the genus. Galls are usually found from June through September. While I have found this species only in the eastern Sierra on interior rose (*Rosa woodsii*) and southern Utah on *Rosa* sp., it has been found throughout the north-central United States and in several other western locations. It is the most common and abundant cynipid gall inducer in Canada. On Prince Edward Island, for example, the common emergent of the galls was the inquiline *Periclistus* sp., while *Pteromalus* sp. was the most common parasitoid.

Plate 286 (left). Tubular galls of
Rhopalomyia audibertiae on black sage.

Plate 287 (right). Tubular gall of
Rhopalomyia salviae on black sage.

Sage Galls

Western sages (*Salvia* spp.) are considered to host several gall organisms, although I have found only two known species to be common in the California area. A third leafy-bract bud gall has been found in southern California and appears to be a new species (not described here). As with many other groups of plants and insects, further study and clarification is needed to gain a clearer picture of the role of gall insects in the sage community.

Reference: Gagné 1989.

SAGE LEAF GALL MIDGE   *Rhopalomyia audibertiae*
Pl. 286

This midge induces tubular, pubescent, nondetachable, thick-walled, monothalamous leaf and petiole galls on both black sage (*Salvia mellifera*) and white sage (*S. apiana*). Galls begin development as early as February. These galls are gray green, sometimes reddish or maroon, have an apical opening, and measure 8 mm wide by up to 12 mm long. Sometimes individual galls coalesce with others to form globular masses that severely distort the leaves. Globular galls may be attacked by inquilines, thus altering the normal tubular shape. Galls protrude on both sides of the leaves, with the dorsal side being a shallow, convex lump and the opposite side looking like a volcanic column with a depression at the tip. Adults emerge through these hair-lined openings at the apex. These spring galls usually produce adults in April and May.

TUBE GALL MIDGE   *Rhopalomyia salviae*
Pl. 287

This midge induces tubular, pubescent, monothalamous galls on the leaves and stems of black sage *(Salvia mellifera)* and perhaps other species of sage. Stem galls appear to be more common than leaf galls. The cylindrical galls project out usually at right angles to the stems and are often found singly just below petioles. Galls measure 5 mm long by 2 mm in diameter and are reddish green. Unlike the galls of the Leaf Gall Midge (*R. audibertiae)*, the galls of this gall midge are straight sided and occur only on one side of the stem. Fresh galls have been found in early February. Larvae pupate inside the galls, with adults emerging through a hole in the gall apex.

Sagebrush Galls

Sagebrush (*Artemisia* spp.) occupies vast areas of coastal, foothill, and arid regions in the West. Unlike sage, this group of shrubs is galled by more than 32 species of gall midges (Cecidomyiidae), two tephritid fruit flies (Tephritidae), and an eriophyid mite (Eriophyidae). Only a few of these are described here. In the case of Great Basin sagebrush *(A. tridentata)*, for example, a large network of gall midges (28 species) and fruit flies and their associated inquilines, parasites, and hyperparasites depend on the presence of this host throughout its range of millions of acres in the West. Most of the galls associated with these shrubs in this guide are covered with three-forked plant hairs. While not as colorful as the galls of blue oaks *(Quercus douglasii)*, the galls of sagebrush are nevertheless intriguing and worthy of continued study.

References: Keifer 1952; Pritchard 1953; Fronk et al. 1964; Force 1974; Jones et al. 1983; Haws et al. 1988; Gagné 1989; Goeden 1990; Hufbauer 2004.

Plate 288. Pit galls of *Aceria paracalifornica* on California sagebrush.

SAGEBRUSH GALL MITE   *Aceria paracalifornica*
Pl. 288

This mite induces swollen, globular pit galls with erineum on the thin, needlelike leaves of California sagebrush *(Artemisia californica)*. These round to irregular-shaped galls occur singly or in clusters in the middle of the thin leaflets, often distorting and bending them. Individual galls measure 2 mm in diameter. The center depression of the pit galls is usually lined with brown hairs. Mites attack the spring needles and remain in their erineum depressions throughout summer and fall. While little is known about the biology of this common mite, it likely overwinters as eggs among bud scales. Severe infestations can stunt plant growth.

STEM GALL TEPHRITID   *Eutreta diana*
Pl. 289

This fruit fly induces spindle-shaped, integral, monothalamous galls in the terminal stem tissues of new shoots of Great Basin sagebrush *(Artemisia tridentata)* and, perhaps, other related species. These pubescent galls usually have leaves emerging from the sides. Galls apparently begin development in fall, remain small during winter, and resume growth and reach maturity in spring. Galls measure 10 to 12 mm long by 8 to 10 mm in diame-

Plate 289. Stem gall of *Eutreta diana* on Great Basin sagebrush.

ter. They are usually gray green and often have a rose purple flush. It has been suggested that certain inquilines and parasitoids influence the development of the red purple tones, even though sun exposure is usually the cause. With age, the galls turn brown. While small amounts of whitish, frasslike material are found, no solid feces are usually present. Sometimes the apical bud of the galled branch is killed by larval activity, and no further stem elongation occurs. Adult emergence occurs in May, and nearly all adults emerge within a week. The dark wings are thought to absorb additional heat, which facilitates early morning activity. As with other tephritids, the courtship and mating activities of these flies are elaborate. Females deposit eggs directly into axillary and terminal buds. The Stem Gall Tephritid usually produces one generation per year. It occurs on *A. filifolia* in Texas, but the life cycle there is different than in California. One study found 147 galls of *Eutreta* on the same shrub. The galls of this fly can also occur on the same shrub along with those of other tephritid fruit flies and gall midges.

Larvae are heavily parasitized by *Eupelmus* sp., *Pteromalus* sp., *Eurytoma* sp., *Rileya* sp., *Tetrastichus* sp., and *Torymus citripes*. The weevil *Apion* sp. is an inquiline. The more common predators of larvae and pupae appear to be insectivorous birds, namely Bushtits *(Psaltriparus minimus)*, that peck holes in the galls to get at the insects.

Plate 290. Leaf gall of *Rhopalomyia calvipomum* on Great Basin sagebrush.

Plate 291. Cluster of galls of *R. calvipomum* involving the entire leaf.

SAGEBRUSH "PLUM" GALL MIDGE

Rhopalomyia calvipomum

Pls. 290, 291

This midge induces globular leaf galls that protrude mostly on the lower surface of the leaves of Great Basin sagebrush (*Artemisia tridentata*). Mature galls hang from the underside of leaves like eggs. The surface is smooth and glabrous. Unlike the galls of the Sponge Gall Midge (*R. pomum*) (soft and spongy), the flesh of this species' gall is relatively firm. Galls measure 9 to 26 mm long by 8 to 20 mm wide. Early galls appear in March and April and are often wine red or violet colored. The basic biology of all the midges in this group is similar, as follows. Larval and pupal development takes place within the galls. Larvae overwinter in the galls and pupate in spring at the surface, with adults emerging by June. While males mate repeatedly, females mate only once and oviposit within a few hours after mating. The bulk of their reproductive activity appears to occur in the morning. Galled leaves often drop by August.

SAGEBRUSH LEAF GALL MIDGE

Rhopalomyia clinata

Pl. 292

This midge induces small, tubular, thin-walled, monothalamous galls on the leaflets of California sagebrush (*Artemisia californica*). These galls protrude at an oblique angle to the leaflet. Galls measure 1 mm in diameter by 2 mm long and are minutely pubescent.

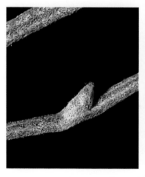

Plate 292. Leaf gall of *Rhopalomyia clinata* on California sagebrush.

Plate 293. The galls of *Rhopalomyia floccosa* on California sagebrush.

Some are pointed at the apex, while others are obtuse. These galls are yellow green or reddish brown. Adults are known to emerge in spring. While the biology of this species is not well known, it is most likely similar to that of other *Rhopalomyia* species.

CALIFORNIA SAGEBRUSH GALL MIDGE *Rhopalomyia floccosa*
Pl. 293

This midge induces white, woolly, monothalamous bud galls on California sagebrush *(Artemisia californica)*. These galls often occur continuously along main stems for several centimeters. Galls measure 5 mm in diameter unless they coalesce, and then a continuous mass may reach 30 to 40 mm long. When the long, white hairs are stripped off the smaller gall bodies, the galls measure 2 mm in diameter. These galls are difficult to detach. Adults have been reared in November. Gall formation, however, does not begin until the following spring.

BEAD GALL MIDGE *Rhopalomyia hirtibulla*
Pl. 294

This midge induces small, pubescent, monothalamous, round to slightly conical galls on the leaves of Great Basin sagebrush

Plate 294. The galls of *Rhopalomyia hirtibulla* on Great Basin sagebrush.

(Artemisia tridentata). Most of these galls protrude on both sides of the leaves and are conical on both ends. Some hang on one side of the host leaf. Galls measure 1.5 mm in diameter by about 1 mm in length. Galls are well developed in May. Larvae can be found at nearly any time of year, although adults emerge the following spring through holes at the top of the galls.

SAGEBRUSH "APPLE" GALL MIDGE   *Rhopalomyia*
Pl. 295   *hirtipomum*

This midge induces round, densely pubescent, solitary, mono-thalamous galls on either surface of the leaves of Great Basin sagebrush *(Artemisia tridentata)*. Galls measure 12 mm in diameter. These galls seem to first appear in late summer or early fall, measuring about 3 mm in diameter, and remain small until the following spring when they resume growth (immature larvae overwinter). By mid-May, galls measure 7 mm in diameter in many areas. Galls reach maturity quickly and begin dropping in July and August. Adults emerge in April. The initial stages of gall formation, however, are slow, with gall development being delayed for as much as four months following oviposition, as in several other species. The weevil *Apion sordidum* is an inquiline in the gall of this species.

Plate 295. The gall of *Rhopalomyia hirtipomum* on Great Basin sagebrush.

Plate 296. The gall of *Rhopalomyia medusirrasa* on Great Basin sagebrush.

WOOLLY BUD GALL MIDGE　　*Rhopalomyia medusirrasa*
Pl. 296

This midge induces globular, leafy-pubescent, polythalamous galls on the buds of Great Basin sagebrush (*Artemisia tridentata*). These large galls are actually composed of numerous leaflike structures that are covered with long, forked hairs, which give the galls an overall white green color. Galls begin development in October, rest during the winter, and reach full size the following spring. These spring galls measure 20 to 25 mm in diameter and contain up to four larvae. Adults emerge in April and May. The larvae, pupae, and adults of this species are basically indistinguishable from those of *R. medusa*. The major differences exist with the galls. The galls of the Woolly Bud Gall Midge have the long white hairs, while the galls of *R. medusa* are hairless.

SPONGE GALL MIDGE　　*Rhopalomyia pomum*
Pl. 297

This midge induces large, round, globular, irregular-shaped, pubescent, soft, spongy galls on Great Basin sagebrush (*Artemisia tridentata*). These collapsible galls are by far the most obvious and common of the midge galls on this sagebrush in the eastern Sierra of central California and throughout the high desert country of the Great Basin. This species is noted for producing two

Plate 297. The gall of *Rhopalomyia pomum* on Great Basin sagebrush.

forms of leaf galls: one is monothalamous and uniform; the second is polythalamous and deeply fissured or multilobed. The number of lobes has been shown not to be associated with the number of gall midge larvae, but instead with the number of parasitoids present (see Hufbauer 2004). Parasitoids may contribute to lobe formation by disrupting chemical cues from the host larvae. Color varies from reddish brown or purple to completely green. Galls occur either midleaf or at the tip and hang from one side of the leaf with a corresponding dimple on the opposite surface. Both forms of galls can occur on the same host shrub. Larval chambers are located near the base of the galls. A single fiber runs from the wall of each chamber to a brown spot on the surface of the galls. What function this serves is not clear. Galls measure up to 45 mm in diameter. Gall development begins in October and follows the pattern of other species, with a winter rest and full growth attainment in spring. Some galls and host leaves shed in August, while others remain on the shrub for over a year. Pupae form partially extended from the surface of the galls. Adult emergence takes place May to June, with females emerging before dawn. Males emerge soon thereafter and collect into large groups that hover over galled shrubs, searching for females. Females mate only once, with oviposition occuring within a few hours after mating. Eggs are deposited on buds. In most cases, individual galls produce either all males or all females, with less than 6 percent of the galls producing both males and females.

The weevil *Apion sordidum* is an inquiline in these galls, and the harvester ant *Pogonomyrmex owyheeli* is a major predator of

Plate 298. The galls of *Rhopalomyia tumidibulla* on Great Basin sagebrush.

emerging adults. Several parasitoids including *Platygaster* sp., *Torymus* sp., and *Tetrastichus* sp. have been reared from these galls, but their relationship to the host is not clear.

SAGEBRUSH BLISTER GALL MIDGE   *Rhopalomyia*
Pl. 298   *tumidibulla*

This midge induces small, integral, lenticular, monothalamous galls in the leaves of Great Basin sagebrush (*Artemisia tridentata*). These small galls can be easily overlooked and measure 3 to 4 mm long by 2 mm wide. The leaf swellings protrude slightly on each side of the galled leaves. Galls are gray green, similar to the host leaves. Galls occur one or two per leaf, located at various points along the length of the leaf. Adults emerge in March and April. The biology of this species is similar to that of the others mentioned.

Saltbush Galls

Over 30 species of saltbush (*Atriplex* spp.), are listed in California. This group of arid region plants is known by a variety of names including shadscale, spearscale, spear oracle, wedgescale, smallscale, and arrowscale. For the most part, the shrubs in this genus play host to either gall midges (Cecidomyiidae) or tephritid fruit flies (Tephritidae) as gall inducers. A rare leaf-mining fly (Agromyzidae) induces galls on two species of saltbush. At least nine species of gall midges are known to gall saltbush in California, the galls of which are described here. It is suspected that saltbush, as a host group, likely has numerous gall inducers that have not yet been described. In one study, researchers found that 12 species of gall inducers on two species of saltbush supported 37 species of parasites, predators, and inquilines. There is always much more to the ecological picture than the human eye sees at a glance.

References: Gordh and Hawkins 1982; Hawkins and Goeden 1982; Gagné and Hawkins 1983; Hawkins and Goeden 1984; Spencer and Hawkins 1984; Hawkins et al. 1986; Hawkins and Unruh 1988; Gagné 1989; Sanver and Hawkins 2000.

SALTBUSH AGROMYZID FLY   *Ophiomyia atriplicis*
Pl. 299

This leaf-mining fly induces leafy, monothalamous, lateral or terminal bud galls on *Atriplex polycarpa* and fourwing saltbush (*A. canescens*). These rather large galls measure up to 30 mm in diameter, but individual galls are usually 10 to 15 mm across. Galls are composed of dense clusters of gray green, elongated, thin leaves that radiate outward. Prior to pupation, each larva scrapes the chamber wall near the apex of its gall, leaving a small, thin area of tissue, like a window. Pupation occurs within the gall, with the adult exiting through the preformed window. Eggs are laid at the base of buds, and the larvae enter the gall after gall development begins. New galls develop between fall and spring. Old galls may persist for a while. Research indicates that this fly may go through two or three generations in a year. Galls that began development in January issued adults in March. A second round of gall development commenced almost immediately, with adults leaving these galls in May. During the following

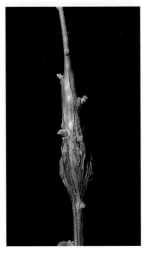

Plate 299. Bud gall of *Ophiomyia atriplicis* on saltbush.

Plate 300. Integral stem gall of *Neolasioptera willistoni* on desert holly.

fall, new galls began development in November, accounting for three generations in 12 months. This species is a rarity among leaf-mining flies, which are generally stem and leaf miners (a few species mine seeds). Only a few species of leaf-mining flies are known as gall inducers worldwide. *Ophiomyia atriplicis* is a southern California gall inducer.

STEM GALL MIDGE   *Neolasioptera willistoni*
Pl. 300

This midge induces integral, monothalamous stem galls on *Atriplex polycarpa*, desert holly *(A. hymenelytra)*, and fourwing saltbush *(A. canescens)*. There are actually two forms of these galls: one form occurs near the tip of fruit-bearing branches, is somewhat fissured, and has an abrupt base; the second form is a smooth, gradually tapered stem swelling. Galls measure 25 to 30 mm long by 5 mm in diameter. Pupation takes place inside the galls. Adults may be found emerging nearly year-round, suggesting that this species may have multiple generations per year. This species occurs from New Mexico to California.

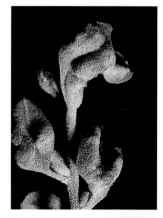

Plate 301. Blister galls of *Asphondylia atriplicicola* on saltbush.

SALTBUSH BLISTER GALL MIDGE   *Asphondylia*
Pl. 301   *atriplicicola*

This midge induces swollen, integral, monothalamous leaf blisters on fourwing saltbush *(Atriplex canescens)*, *A. spinifera*, and possibly other species. One to several galls may occur on a single leaf, sometimes distorting the leaves. Galls measure 4 mm long by 2 mm wide and are smooth with appressed small hairs. These galls occur along the midrib or the leaf margins. Some galls also occur on the petiole, bending it radically. This gall midge can be found in nearly all stages of its cycle year-round. It has been recorded from New Mexico to California.

TUMOR STEM GALL MIDGE   *Asphondylia atriplicis*
Pl. 302

This midge induces large, smooth, soft, fleshy galls on fourwing saltbush *(Atriplex canescens)* and *A. confertifolia*. These large, round-globular, sometimes potato-shaped galls measure from 7 to 15 mm in diameter and up to 50 mm long. Although the surface of the galls is relatively smooth and shiny, there are randomly scattered, shallow pits and whitish blotches. Galls vary from olive green to red. As many as 70 larvae have been found in one large gall mass. The larval chambers are lined with a white fungus. Because the larval chamber walls lack the nutritive layer characteristic of cynipid galls, it is thought that the larvae of all *Asphon-*

Plate 302. Integral stem gall of *Asphondylia atriplicis* on fourwing saltbush.

Plate 303. Galls of *Asphondylia caudicis* on fourwing saltbush occurring next to the larger gall of *A. atriplicis*.

dylias, as well as other generic groups, feed on the fungus or fungal by-products (see the discussion of ambrosia galls in the section on gall midges). While this relationship seems to be obligatory, it is not yet clear how the fungus is transported to new galls. As if things were not complicated enough, there is also an inquilinous eulophid, *Tetrastichus cecidobroter,* whose larvae induce endogalls within the galls of this midge species. While the inquiline does not prey on the gall midge directly, the midge larvae are often crushed by the developing endogalls (see the Bubble Gall Tephritid *[Aciurina trixa]* account and pl. 264.

SHADSCALE BUD GALL MIDGE   *Asphondylia caudicis*
Pl. 303

This midge induces round to oval, gray green, monothalamous bud galls on fourwing saltbush *(Atriplex canescens)* and possibly other species. Galls measure 4 mm in diameter and have a short pile pubescence. They are usually found at the base of petioles, but I have found a few attached to the base of the galls of the Tumor Stem Gall Midge *(Asphondylia atriplicis).* Adults emerge by March. It has been found from New Mexico to California. Little else is known about this species.

Plate 304. Fluffy gall of *Asphondylia floccosa* on saltbush.

Plate 305. Stem galls of *Asphondylia nodula* on desert holly.

WOOLLY STEM GALL MIDGE   *Asphondylia floccosa*
Pl. 304

This midge induces large, white, cottonlike gall masses on *Atriplex polycarpa, A. spinifera,* and *A. confertifolia.* These galls can be monothalamous or polythalamous. Galls are often round to oblong and usually measure about 30 mm long by 15 mm in diameter. Galls appear during two distinct periods of the year: a spring period from December to June, and a fall period from July to December. Newly developing galls can be found just about any time of year, depending on locality. Small (less than 15 mm) summer galls form near the top of the host shrub, while larger galls (more than 30 mm), forming in fall or winter, occur farther down among the branches. Galls cover most of one side of the branch and sometimes completely encircle it. The white hairs are tightly compressed and often have a pink or rose tinge near the base, particularly in young specimens. Pupation occurs at the surface of the galls, with the black pupal cases left behind after the adults emerge. This midge likely has several generations per year. This species has been collected from the Colorado and Mojave Deserts of California to the Sonoran Desert of Arizona.

Plate 306. Fall phase of *Asphondylia* sp. on desert holly.

Plate 307. Spring phase of *Asphondylia* sp. on desert holly.

NODULAR STEM GALL MIDGE   *Asphondylia nodula*
Pl. 305

This midge induces rough, fissured, integral stem galls on *Atriplex polycarpa* and desert holly *(A. hymenelytra)*. These galls appear as eruptions of the stem and may be round to elongated and the same color. Small galls measure 4 to 10 mm in diameter, but when galls coalesce they can be significantly larger. These larger masses can contain hundreds of larvae. Galls begin appearing in January, with the continuing development of new galls through March. While there are some general behaviors that are characteristic of *Asphondylia* spp., the exact biology of this species is not well known.

POM-POM BUD GALL MIDGE   *Asphondylia* sp.
Pls. 306, 307

This midge induces bristly, bright pink, round, monothalamous bud galls on desert holly *(Atriplex hymenelytra)* in fall. Some galls are a dusty pink, while others have white-tipped red hairs or are reddish pink and resemble bright pom-poms. The hairs are dense and compact and do not rub off easily. Galls measure 10 mm in diameter. These nondetachable galls begin development in fall but do not reach full size until spring. In spring, galls are creamy white and measure 25 mm long by 15 mm wide and high. The central larval chamber is surrounded by bright green flesh. The walls of the chamber are smooth except for a patch of white fungus at one end. Young larvae are found in fall and reach full size by March. Adult emergence likely takes place in late summer or fall.

Scale Broom Galls

Scale broom *(Lepidospartum squamatum)* is a common shrub in the Mojave Desert and southwestern California and extends north to southern Alameda County. In some areas it is considered to be a noxious and invasive species. Like other plants, however, it has its place in the ecosystem. Two little-known species of arthropods utilize scale broom as a host, resulting in noticeable galls.

References: Keifer et al. 1982; Miller 2005.

SCALE BROOM GALL MITE   *Eriophyes lepidosparti*
Pl. 308
This mite induces compact clusters of expanded buds along the branches of scale broom. In spring, when these galls are fresh, they look like clusters of tiny, green cabbages. Each individual round gall measures 4 mm in diameter. Clusters can reach 12 to 15 mm across. Individual galls are composed of a small rosette of leafy bud scales. New galls often develop at the sites of old galls. These galls and their mites tend to have a stunting effect on new growth. Some have numerous stunted shoots emerging from the galls. The mites live among the bud scales.

SCALE BROOM GALL MOTH   *Scrobipalpopsis* sp.
Pl. 309
This moth induces gently tapered, elliptical, integral stem galls on scale broom. These smooth, glabrous galls are fresh in March and April, with moth caterpillars 3 to 4 mm long. Galls measure 20 mm long by 5 mm in diameter. Most of the galls are green except for the apex, which turns brown. There is no growth beyond the gall. Caterpillars pack frass in the top of the galls, which may be related to the color change and the prevention of growth beyond the gall tips. Pupation occurs in the galls. Galls collected in

Plate 308. Bud gall of *Eriophyes lepidosparti* on scale broom.

Plate 309. Integral stem gall of *Scrobipalpopsis* sp. on scale broom.

April issued adults in early June. Little else is known about the biology of this new species of moth discovered during the field-work for this guide.

Service-berry Galls

The toothy-edged, dull green, rounded leaves of service-berry (*Amelanchier* spp.) distinguish this shrub in its mountainous habitat among pines (*Pinus* spp.), incense-cedars *(Calocedrus decurrens)*, and other conifers and shrubs. For reasons not well understood, service-berry has become host to a diverse array of related gall midges whose taxonomy is challenging. The midge galls described here are currently considered to belong in the genus *Blaesodiplosis.* Until further research clarifies the taxonomy of this complex group, they are assigned an alphanumeric label for each gall type. In addition, this shrub supports a sac fungus and an aphid (Aphidae) that induce galls.

References: Sinclair et al. 1987; Gagné 1989.

Plate 310 (upper left). Witches' broom caused by *Taphrina amelanchierii* on service-berry.

Plate 311 (upper right). The galls of the Leaf-Roll Aphid on service-berry.

Plate 312 (left). The galls of *Blaesodiplosis* sp. A on service-berry.

SERVICE-BERRY FUNGUS   *Taphrina amelanchierii*
Pl. 310

This fungus induces compact, dense witches' brooms at the tips of shoots of *Amelanchier alnifolia* and Utah service-berry (*A. utahensis*) in spring. These witches' brooms are quite noticeable because of the dense collection of short branches and the associated swelling in the immediate area. These brooms can easily exceed 20 cm across. For a complete discussion of the biology of these fungi, see the section on "sac fungi." These massive, congested collections of branches serve as excellent hiding and feeding places for a broad variety of spiders, scale insects, and other secondary invaders.

LEAF-ROLL APHID   *Undescribed*
Pl. 311

This aphid induces roll galls on the terminal leaves of service-berry. Both edges of the leaves are usually rolled up and inward. More often than not, all of the terminal leaves of a particular spring shoot are affected by this aphid. Roll galls measure 20 mm long by 3 to 4 mm in diameter. Leaf tissues are swollen and distended. Large, black ants are often seen walking over these galls. Many leaf-rolling aphids have alternate hosts. The biology of this new species and any potential alternate host in the Sierra Nevada are unknown.

WOOLLY LEAF GALL MIDGE   *Blaesodiplosis* sp. A
Pl. 312

This midge induces hairy, white, triangular-shaped, monothalamous galls on the upper surface of the leaves of both *Amelanchier alnifolia* and Utah service-berry *(A. utahensis)*. These red-tipped galls are usually in such close proximity to each other, they often appear as a single, hairy, white mass. Galls occur near the midrib vein at the base of the leaf and petiole, sometimes incorporating the entire leaf. Individual galls measure 6 mm high by 3 to 4 mm wide at the base. There is a slight protrusion on the corresponding, lower side of the leaf. Given the snow-covered winters throughout much of the range of the host plants, this midge most likely has one annual generation limited to late spring and summer.

TOOTH GALL MIDGE   *Blaesodiplosis* sp. B
Pls. 313, 314

This midge induces glabrous, triangular, monothalamous, flat-sided galls that hang from the underside of the leaves of both species of service-berry. The galls are characterized by a corresponding slit or opening on the dorsal side of the galled leaf that is usually red. These galls are usually yellow beige at the base near the leaf, but bright red at the narrowed tip. Individual galls measure 4 mm high by 2 mm wide at the base. The galls occur singly between the lateral veins. Sometimes both the Woolly Leaf Gall Midge *(Blaesodiplosis* sp. A) and the Tooth Gall Midge can be found on the same leaf.

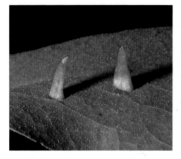

Plate 313. Nail galls of *Blaesodiplosis* sp. B on the dorsal surface of a service-berry leaf.

Plate 314. Nail galls of *Blaesodiplosis* sp. B on the ventral surface of a service-berry leaf.

RED KNOB GALL MIDGE *Blaesodiplosis* **sp. C**
Pls. 315, 316

This midge induces columnar, round-topped, knobby galls on the dorsal surface of the leaves of service-berry in the spring. The glabrous galls are recessed on the underside of the leaf, where the slit opening shows. These integral galls are usually bright wine red on the dorsal side and light green on the underside. Galls measure 2 to 3 mm high by 1 to 2 mm wide. A top-down view reveals an oblong shape that reaches 2 to 3 mm long. These galls occur in large numbers per leaf and often flank both sides of the midrib vein. They can occur along with all other species of *Blaesodiplosis* on the same host leaf. This new species was discovered in 2004 during the preparation of this guide.

PURSE GALL MIDGE *Blaesodiplosis* **sp. D**
Pls. 317, 318

This midge induces small, lumpy, reddish galls on the dorsal surface of service-berry leaves. Each of these monothalamous galls is marked with a white-haired slit on the dorsal surface. The portion of the galls that protrudes from the underside of the leaves is covered with short, white hairs. These round galls measure 1 to 2 mm high and wide. Some forms are oblong and may reach 3 mm long by 1 mm wide. These conspicuous galls usually occur in

Plate 315. Dorsal view of the galls of *Blaesodiplosis* sp. C on service-berry.

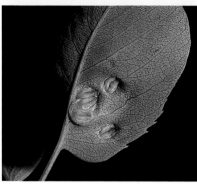

Plate 316. Ventral view of the galls of *Blaesodiplosis* sp. C showing the slit openings.

large numbers per leaf and often in tight clusters. A similar species has been recorded on the East Coast. This new species was discovered in 2004 during the fieldwork for this guide.

RED LIP GALL MIDGE   *Blaesodiplosis* **sp. E**
Pls. 319, 320
This midge induces a monothalamous, glabrous, oblong leaf gall similar to that of the Purse Gall Midge (*Blaesodiplosis* sp. D). These red-lipped galls have a noticeable slit on the dorsal surface of the leaf and are 1 mm high by 2 mm long. Galls protrude from the underside in a bulging, smooth, light green, glabrous knob form. This latter feature distinguishes these galls from those of the Purse Gall Midge, which are hairy-white on the underside. In

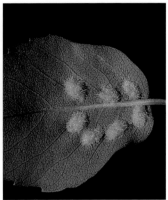

Plate 317. Dorsal view of *Blaesodiplosis* sp. D showing the slit opening.

Plate 318. Ventral view of the galls of *Blaesodiplosis* sp. D.

Plate 319. The galls of *Blaesodiplosis* sp. E (left side) with slits.

Plate 320. Opposite side of *Blaesodiplosis* sp. E.

the Sierra, galls are well formed by mid-June. Little is known about the biology of these midges. While this new species was also discovered during the preparation of this guide, its existence points to the complexity of the taxonomy, the need for further research, and the diversity of species that are little known.

Plate 321. Flower gall of *Asphondylia garryae* on silk tassel.

Silk Tassel Galls

Silk tassel (*Garrya* spp.) occurs in coastal and montane environments. Apparently, only one gall organism, a gall midge (Cecidomyiidae), is associated with these shrubs. The galls of this midge are found on two different plant organs.

References: Russo 1979; Gagné 1989.

SILK TASSEL GALL MIDGE　　　　*Asphondylia garryae*
Pl. 321

This midge induces round-globular, fleshy, monothalamous galls on axillary buds and catkins of *Garrya buxifolia* and *G. fremontii*. Galls are composed of greatly expanded, overlapping bud scales. Galls measure 18 mm long by 14 mm in diameter, and the central larval chamber measures about 3 mm in diameter. The flesh surrounding the larval chamber is soft and spongy. Galls bear a sparse arrangement of tiny hairs across the surface. Galls are green, red, or purple. They occur singly in axillary buds, but one or more may be found on the tassel-like catkins. Some catkin galls coalesce to form larger galls. The galls of this midge develop during early summer, and by July they are mature. Adults emerge the following spring.

Plate 322. Roll gall of *Phyllocoptes triacis* on snowberry.

Snowberry Galls

Snowberry (*Symphoricarpos* spp.) is one of the most common shrubs of woodlands, pine forests, and roadsides of central to northern California and Oregon. Its distinctive white, fleshy berries help identify this shrub. At least four species of snowberry occur along the Pacific Coast. Snowberry supports three primary gall inducers in the California region: an eriophyid mite (Eriophyidae), a gall midge (Cecidomyiidae), and a sawfly (Tenthredinidae).

References: Keifer et al. 1982; Larew and Capizzi 1983; Gagné 1989.

SNOWBERRY MITE   ***Phyllocoptes triacis***
Pl. 322

This mite induces fleshy, swollen roll galls along the margins of the leaves of snowberry (*Symphoricarpos albus*) and, presumably, other species. The affected leaf margins roll from the undersurface in toward the midrib of the dorsal side. The thick roll galls are hard and usually a lighter green than the dorsal surface of the leaves. Galls can incorporate a portion or all of the affected leaf margins. Galls can measure 10 to 20 mm long by 1 to 2 mm in diameter. Heavy galling by this mite may deform host leaves, but there appears to be no serious damage. This mite has been found galling snowberry in Alaska, Montana, Washington, Oregon, and California. Little is known about the biology of this species.

Plate 323. Roll gall of *"Dasineura"* sp. on snowberry.

SNOWBERRY GALL MIDGE   *"Dasineura"* sp.

Pl. 323

This midge induces compact, linear, glossy, smooth, polythalamous roll galls along the leaf margins of snowberry *(Symphoricarpos albus)* and creeping snowberry *(S. mollis)*. These glabrous galls appear in spring and continue to support active orange larvae well into October. They can be green or reddish purple. Galls roll in from the edge toward the midrib of the upper leaf surface. These galls are very hard and somewhat brittle. Galls measure 20 mm long by 2 to 3 mm in diameter. In some cases, the rolls extend from the basal attachment to the petiole all the way to the tip of the leaf. Up to four orange larvae can be found in each roll gall. Larvae drop to the ground for pupation. While some adults have been reared in August, specimens found in October still had active larvae, suggesting a later emergence for adults. This midge has been found in Oregon and California. The gall was originally thought to be caused by *Resseliella californica,* but recent research has determined that the larvae found in another collection of these galls lack large posterior hooks, which is characteristic of *Dasineura* sp. The larvae I have taken from these galls in October and those examined by others in British Columbia have also proven to be *Dasineura* sp. It may turn out that there are two midges associated with these galls: one the gall inducer, the other an inquiline. This gall is assigned here to *Dasineura* provisionally

Plate 324. Terminal bud gall of *Blennogeneris spissipes* on snowberry.

until it is determined which midge is the inquiline and which is the gall inducer. These gall midges are locally abundant in California, Oregon, and Washington.

SNOWBERRY SAWFLY   *Blennogeneris spissipes*
Pl. 324

This sawfly induces large, fleshy, globular, monothalamous bud galls on snowberry *(Symphoricarpos albus)* and creeping snowberry *(S. mollis)*. The galls of this sawfly vary greatly in size from one region to another. Galls examined in the San Francisco Bay Area measured 10 to 20 mm in diameter, while galls found on creeping snowberry in the Sierra were 35 mm across. These light green galls can be round or multilobed when individual galls coalesce. Some galls have the edges of leaves protruding from the gall mass. Larval chambers are large and usually filled with frass by late July. Galls that form on *S. albus* are less vulnerable to predators than the galls forming on creeping snowberry in the Sierra. In one area, I found more than half of the galls on creeping snowberry had been gnawed open by chipmunks *(Neotamias* spp.) and Golden-mantled Ground Squirrels *(Spermophilus lateralis)* to get at the larvae and pupae. Given the accessibility of available protein, these rodents may be major predators. Little else is known about the biology of this interesting sawfly.

Plate 325. Integral stem gall of *Diastrophus kincaidii* on thimbleberry.

Thimbleberry Galls

Thimbleberry *(Rubus parviflorus)* is common in shaded environments throughout the Pacific Coast all the way into southeastern Alaska. While other members of the genus, particularly in the East, host a variety of gall organisms, thimbleberry supports only one known gall inducer, a bisexual cynipid wasp (Cynipidae).

References: Weld 1957; Wangberg 1975.

THIMBLEBERRY STEM GALL WASP   *Diastrophus*
Pl. 325   *kincaidii*

This wasp induces large, globular, polythalamous, knobby stem galls that often bend the branch. These galls occur on new spring growth, reaching 80 mm long by 35 mm in diameter. They often have short bristles or spiny hairs that are pliable when fresh, but brittle by fall. When fresh, the galls are green to reddish brown. Some galls are round while others are potato shaped with numerous knobs. The number and location of larvae have a major influence on the length and shape of the galls. Larval chambers are arranged radially around a thick central core. Occasionally, larvae chew through the walls of their particular chambers, breaking into the larval cavities of others. They apparently coexist peacefully in a common chamber, as two pupae have been found sharing the same space. Galls first begin to show about 10 days after egg deposition. The larvae enter diapause in fall, with adults emerging the following spring. Studies show that males emerge before females. There appears to be more walking about and little flying for the adults following emergence. After mating, females deposit several eggs close together into the new, soft spring growth. Wangberg (1975) found that several females deposit eggs close together, and the resulting larvae are incorporated within one large gall. Populations of this gall wasp are extremely localized, with hundreds of galls occurring in one area but completely absent from another area where the host plant flourishes. This species has been found in California and Oregon but can be expected elsewhere in the range of its host. At least nine species of parasitic wasps and one weevil, as an inquiline, have been associated with the galls of this cynipid.

MISCELLANEOUS GALLS

Native Plant Galls

Several plants do not fit into the previous two categories (trees and shrubs) that support known gall organisms but are worth mentioning. Among them are ferns, grasses, violets, lilies, and wild grapes. Each of these plants supports a unique gall organism. Note that Phillip Munz (1963) classified the lilies mentioned in the following species discussion in the genus *Brodiaea*. With the revision of *The Jepson Manual: Higher Plants of California* (Hickman 1993), the genus *Brodiaea* was split into three different genera: *Brodiaea, Triteleia,* and *Dichelostemma.*

After the native plants are brief highlights of known galls on ornamental plants. As throughout this guide, keep in mind many more galls are out there than can be included here.

GRAPE LEAF MITE   *Colomerus vitis*
Pl. 326
This mite induces erineum galls on the leaves of wild grapes *(Vitis californica)* and the cultivated grape *(V. vinifera)*. When these mites gall the leaves, dozens of the lumpy, blisterlike, green to red, irregularly shaped galls can be seen on the surface of the leaves. Each lump has a 2 mm deep corresponding depression on the underside of the leaf, comprising the hair-lined erineum pocket. The hairs can be beige, pink, rose, or fuchsia colored. On the upper surface, galls measure 5 to 10 mm in diameter. Sometimes, galls coalesce and appear larger. The three forms of this species each cause a different kind of damage. In addition to the leaf erineum form, another form of this mite attacks the flower buds, preventing flowering and grape development. A third form of this mite attacks young leaves, causing leaf-curl. This species is widespread.
Reference: Keifer et al. 1982.

Plate 326 (right). Bead galls of *Colomerus vitis* on wild grape.

Plate 327 (below). Galls of *Edestochilus allioides* on knot grass.

GRASS GALL MIDGE   *Edestochilus allioides*
Pl. 327

This midge induces large, globular, integral galls on stems of knot grass *(Paspalum distichum)*. In spring, these galls are fleshy, green, and about 10 to 15 mm in diameter. The outside of the galls is covered with overlapping leafy bracts. Once this seasonal grass dries, the galls become extremely hard. Larvae remain in the galls and overwinter in diapause until the next spring, when they pupate. Adults lay their eggs near the tip of grass stems, where the galls ultimately develop. Grass galls are relatively unknown since little research has been done in this area. However, many

Plate 328. Stem gall of *Lasioptera* sp. on Ithuriel's spear.

species of grass galls are out there, and with more time and curious naturalists, our knowledge and the number of identified species of grass-galling midges will certainly grow.

Reference: Gagné 1989; Harris et al. 2003.

LILY STEM GALL MIDGE   *Lasioptera* sp.
Pl. 328

This midge induces tapered or abrupt, integral, polythalamous stem galls just below the flowers of Ithuriel's spear *(Triteleia laxa)*. This gall inducer has also been recorded on *Brodiaea californica* var. *leptandra* in Napa County. Even though I have found blue dicks *(Dichelostemma capitatum)* and ookow *(D. congestum)* growing right next to galled Ithuriel's spear that were unaffected, this midge may gall other related species. Galls can be smooth and symmetrically tapered or abrupt and somewhat knobby. They measure to 60 mm long by 15 mm in diameter. These spring galls stand out because of their size and are usually locally abundant. The central larval chamber is large and usually filled with a dozen or more orange larvae. The larvae pupate inside the galls and emerge in spring as adults along with the development of a new season's growth of Ithuriel's spear and other related lilies. Little else is known about the biology of this species.

References: Munz 1963; Gagné 1989; Hickman 1993.

VIOLET GALL MIDGE   *Prodiplosis violicola*
Pl. 329

This midge induces a fleshy, swollen roll gall along the margins of the leaves of violets (*Viola* spp). These galls occur on numerous species and varieties of both native and ornamental violets. In response to the larvae, the edges of the leaves roll in toward the midrib of the upper surface of the leaves. One or both margins can be rolled inward, distorting the affected leaf. Galls measure 25 to 30 mm long by 5 to 7 mm in diameter. Eggs are laid while leaves are still in bud form. Full-grown larvae drop to the ground, where they pupate. Several generations per year can occur. This midge was once regarded as a major pest of ornamental violets. Although it can severely distort the host plant, it does not kill the perennial hosts. It has been found in several locations in the East, as well as here in the West.

Reference: Gagné 1989.

VIRGIN'S BOWER GALL RUST   *Puccinia recondita*
Pl. 330

This fungus induces deep pocket galls on the leaves of pipestems (*Clematis lasiantha*) and perhaps other virgin's bower species (*Clematis* spp.). These pockets are noticeable depressions on the upper surface of leaves, with corresponding bulges on the lower surface. In March, the inner lining of the depression is smooth and bright orange, while the bulge on the lower leaf surface is covered with small, orange bumps or pustules. Galls measure 12 mm across by 5 mm deep. There is usually one pocket gall per leaf. The rust pustules that break through the epidermis of the lower surface of the leaves increase water transpiration. This fungus is found worldwide. Apparently several varieties of this fungus affect different host plants including columbine (*Aquilegia* spp.), larkspur (*Delphinium* spp.), and buttercups (*Ranunculus* spp.). Some of these rust varieties use wheat and wild rye-

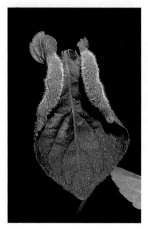

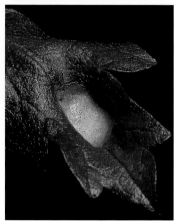

Plate 329. Leaf roll galls of *Prodiplosis violicola* on hybrid violet.

Plate 330. Dorsal view of the pocket gall caused by *Puccinia recondita* on virgin's bower.

Plate 331. Gall of *Taphrina californica* on wood fern.

grasses as alternate hosts. I have seen this fungus only in the southern Sierra foothills, but it may occur elsewhere in the range of its host.

References: Westcott 1971.

WOOD FERN SAC FUNGUS

Taphrina californica

Pl. 331

This fungus induces green, fleshy, gelatinous, swollen galls at the tip of pinnules of wood fern *(Dryopteris arguta)*. These galls are succulent and emerald green. Galls measure 10 to 20 mm wide by 5 mm thick. They are convex on one side, concave on the other. Galls form along the margins of the pinnules but can also encompass leaf tissue to the midrib and beyond. Galls originate in the epidermis of the fern. The fungal hyphae actually develop in the outer walls of the surface cells. These spring galls are quite noticeable once you see the first one. Usually by fall, the galls have dried and turned brown. For a full discussion of the biology of *Taphrina,* see the section on "sac fungi." These galls have been found in Mexico, California, and Oregon. A related species, *T. faulliana,* has been reported on western sword fern *(Polystichum munitum)* in Oregon and Washington. It may also occur in California.

Reference: Mix 1949.

Ornamental Plant Galls

Several galls and gall organisms occur on ornamental plants used in private gardens and public landscaping (pls. 332 to 336). No attempt will be made here to describe these species in detail as with native species previously mentioned. Table 15 gives a general introduction to some of the more common ornamental gall organisms. Further information on these species can be obtained from local Agricultural Pest Advisor offices or several of the cited references. Additionally, several galls occur on exotic weeds that have been introduced into the West. One that is included here is an integral stem gall induced by the cynipid wasp *Aylax hypochaeridis* on the lawn and wayside weed rough cat's ear *(Hypochaeris radicata)*.

References: Keifer et al. 1982; Sinclair et al. 1987; Johnson and Lyon 1991.

TABLE 15 Galls on Ornamental Plants

Host Plant	Agent	Type of Gall
Australian brush-cherry *(Eugenia myrtifolia)*	*Trioza eugeniae*	Psyllid, pit gall (pl. 336)
Cat's-ear *(Hypochaeris* spp.)	*Aylax hypochaeridis*	Cynipid, peduncle swelling
Cork oak *(Quercus suber)*	*Plagiotrochus amenti*	Cynipid, stem gall
Broom *(Cytisus* spp.)	*Aceria genistae*	Mite, globular stem gall
Fruit trees (several) base of tree	*Agrobacterium tumefaciens*	Bacteria, crown gall
Hibiscus *(Hibiscus* spp.) (Hawaii)	*Aceria hibisci*	Mite, bead gall on leaf (pl. 333)
Honey locust *(Gleditsia triacanthos)*	*Dasineura gleditchiae*	Midge, roll gall of leaf
Linden *(Tilia* spp.)	*Phytoptus tilliae*	Mite, nail gall on leaf
Oleander *(Nerium oleander)*, olive *(Olea europaea)*	*Pseudomonas syringae*	Bacteria, knot gall on stem (pl. 334)
Ornamental pepper tree *(Schinus molle)*	*Calophya rubra*	Psyllid, pit gall on leaf (pl. 335)
Peach, nectarine *(Prunus* spp.)	*Taphrina deformans*	Fungus, leaf-curl (pl. 332)

Plate 332. Leaf-curl gall of *Taphrina deformans* on peach.

Plate 333. Bead galls of *Aceria hibisci* on ornamental hibiscus.

Plate 334 (upper left). Stem galls of *Pseudomonas syringae* on oleander.

Plate 335 (upper right). Pit galls of the psyllid *Calophya rubra* on ornamental pepper tree.

Plate 336 (left). Pit galls of the psyllid *Trioza eugeniae* on Australian brush-cherry.

EPILOGUE

A vast number of little known galls and gall-inducing organisms occur in the western United States. Even for identifiable species, we have only limited information available. Add to this all of the species that have yet to be studied and classified, and it becomes clear that there is so much more to learn about these fascinating creatures.

Throughout the 35 years I have been collecting and studying plant galls, I have found many new species, and I still continue to find new ones to this day. I am sure you will too, if you look carefully. The discovery of new species of galls, mushrooms, wildflowers, or any of the other treasures that adorn our planet is that intangible element that fuels our hope, drives our inspiration, and defines our humanity. Let yourself go and embrace nature as your very own treasure trove. Enjoy!

GLOSSARY

Adelgids Insects related to aphids but in the family Phylloxeridae.

Aecial stage The stage in the rust fungus life cycle that produces cuplike or blisterlike asexual fruiting structures with light yellow to orange aeciospores, which are wind dispersed.

Agamic Refers to the asexual or unisexual generation of females only, which reproduce parthenogenetically (without sexual union).

Alate An aphid in the winged stage of its life cycle.

Alternation of generations A reproductive cycle that involves alternating between a spring bisexual generation and a summer-fall unisexual generation, typical of cynipid wasps and some aphids.

Ambrosia Originally, the fungi eaten by certain beetles. In this guide, it is used as an adjective that indicates which gall midges are associated with a fungus. These fungi line the walls of larval chambers. The midges eat either the fungi or the by-products of the fungi.

Bisexual Refers to a generation that has males and females.

Catkin A long floral spike of clustered flowers typical of oaks, willows, alders, and cottonwoods, among others.

Cecidology The study of gall organisms and their interactions with host plants.

Convoluted Refers to tissues that are rolled and folded, or brainlike.

Crenate Refers to a margin with rounded teeth.

Cryoprotectant A chemical or combination of chemicals developed by some insects to prevent freezing.

Diapause A state of dormancy that interrupts complete development in a life cycle, or a period when growth is arrested and metabolism is low.

Dorsal Refers to the upper surface.

Endogall A gall that develops within another gall in response to an invader.

Erineum The hairy, velvety, or pilelike depressions on the surfaces of leaves caused by eriophyid mites.

Fasciation An abnormal enlargement or flattening of plant stems or flower stalks.

Frass The solid fecal material produced by feeding insects and composed of digested plant material.

Frugivorous Refers to feeding on fruit, generally in reference to tephritid fruit flies.

Fundatrix A female aphid that initiates gall formation; also called a "stem mother."

Gall An abnormal plant growth, swelling, or tumor induced by another organism, including fruit and buds altered or aborted by gall organisms.

Gall inducer The organism responsible for initiating gall development.

Glabrous Refers to the absence of hairs.

Heterogeny The reproductive process that involves an alternation of a spring sexual generation with a summer-fall unisexual generation, typical of many cynipid wasps and some aphids.

Honeydew A sweet liquid material produced by some aphids and scale insects as a by-product of their feeding. Honeydew also refers to the sugary material that accumulates on the surface of certain cynipid wasp galls.

Hyperparasite A parasite that specializes in attacking other parasites.

Hypha An individual fungal strand that forms after spores germinate. The plural is *hyphae.*

Inquiline An insect that specializes in eating the plant tissue of a gall, which is induced by another insect. Inquilines often kill and eat other insects confronted inside the gall.

Instar The stage of a larval insect between molts. Gall midges usually have three, while cynipid wasps have five.

Integral Refers to a nondetachable bulge or swelling of a stem, petiole, or leaf.

Larva An immature stage in the life cycle of an insect between the egg and the adult form referred to as caterpillars, maggots, or grubs. The plural is *larvae.*

Leaf Curl Refers to a leaf gall where the tissue is swollen, distorted, and discolored, usually in response to the invasion of a fungus.

Lenticel A wartlike scale or tubercle found on the surface of some galls.

Meristematic tissue Tissue with cells capable of frequent division and responsible for the first phases of growth.

Midrib The central vein of a leaf.

Monothalamous Refers to a gall inhabited by one or more larvae in a common chamber or cavity.

Nonfrugivorous Generally refers to non-fruit-eating, gall-inducing tephritid fruit flies.

Oviposition The act of a female depositing eggs.

Ovipositor The external, egg-laying, tubular apparatus of a female insect.

Parasite An animal or plant that lives off the tissues of another living organism during all or part of its life cycle.

Parenchyma tissues Tissue with thin-walled cells that often store food.

Parthenogenesis Reproduction by a female (agamic) without fertilization or genetic exchange with a male.

Petiole The slender stem that supports the blade of a leaf.

Polythalamous Refers to a gall that contains two or more larvae each in separate chambers or cavities.

Pubescence A covering of soft hairs.

Pupa The transitional stage between larva and adult in insects, with complete metamorphosis. The plural is *pupae.*

Sessile Refers to a leaf without a petiole or stalk.

Spermatophore A slender stalk bearing a sperm sac at the apex. In eriophyid mites, the females carry spermatophores through the winter after contact with males.

Stem Mother In aphids, the female that initiates gall formation and the production of several generations of offspring.

Stylet A needlelike, piercing-sucking structure in the mouthparts of an insect or mite.

Telial stage The sexual stage in the rust fungus life cycle, which comes after the uredinial stage and produces teliospores of various colors that are dispersed short distances, generally not by wind. Many rust fungi survive winter as teliospores.

Tubercle A small tuberlike prominence or nodule.

Umbo A conical projection or small bump.

Unisexual Refers to the stage in the life cycle of an insect (generally cynipid wasps) composed of females only.

Uredinial stage The stage in the rust fungus life cycle that produces asexual blisterlike pustules with orange urediniospores, which are wind dispersed.

Ventral Refers to the underside of a leaf or other body.

Witches' Broom An abnormal cluster of shoots or twigs emanating from a common focal point on stems and branches, caused by either mechanical injury (not a gall) or by the invasion of an organism (a true gall).

REFERENCES

Abrahamson, Warren G., and Arthur E. Weis. 1997. *Evolutionary ecology across three trophic levels: Goldenrods, gallmakers, and natural enemies.* Princeton, NJ: Princeton University Press.

Alleyne, E.H., and F.O. Morrison. 1977. Some Canadian poplar aphid galls. *Can. Entomol.* 109 (3): 321–328.

Al-Saffar, Zuhair Y., and John C. Aldrich. 1998. *Pontania proxima* (Tenthredinidae: Hymenoptera): Natural enemies and defensive behavior against *Pnigalio nemati* (Eulophidae: Hymenoptera). *Ann. Entomol. Soc. Am.* 91 (6): 858–862.

Arthur, Joseph Charles. 1962. *Manual of the rusts in the United States and Canada.* New York: Hafner Publishing.

Bagatto, G., and J.D. Shorthouse. 1991. Accumulation of copper and nickel in plant tissues and an insect gall of lowbush blueberry, *Vaccinium angustifolium,* near an ore smelter at Sudbury, Ontario, Canada. *Can. J. Bot.* 69 (7): 1483–1490.

Bagatto, G., T.J. Zmijowskyj, and J.D. Shorthouse. 1991. Galls induced by *Diplolepis spinosa* influence distribution of mineral nutrients in the shrub rose. *HortScience* 26 (10): 1283–1284.

Balls, Edward K. 1962. *Early uses of California plants.* Berkeley and Los Angeles: University of California Press.

Barrett, S.A., and E.W. Gifford. 1933. Miwok material culture. *Bull. Milwaukee Public Museum Yosemite Nat'l History Assoc.* 2 (4): 192.

Bean, Lowell John, and Katherine Siva Saubel. 1972. *Temalpakh-Cahuilla Indian knowledge and usage of plants,* 123–129. Banning, CA: Malki Museum Press.

Blackman, R.L., and V.F. Eastop. 1994. *Aphids on the world's trees.* Wallingford, UK: CAB International and The Natural History Museum.

Briggs, C.J. 1993. The effect of multiple parasitoid species on the gall-forming midge *Rhopalomyia californica*. Ph.D. thesis. University of California, Santa Barbara.

Briggs, C.J., and J. Latto. 1996. The window of vulnerability and its effect on relative parasitoid abundance. *Ecol. Entomol.* 21 (2): 128–140.

Brooks, Scott E., and Joseph D. Shorthouse. 1997. Biology of the rose stem galler *Diplolepis nodulosa* (Hymenoptera: Cynipidae) and its associated component community in central Ontario. *Can. Entomol.* 129:1121–1140.

Brooks, Scott E., and Joseph D. Shorthouse. 1998a. Development morphology of stem galls of *Diplolepis nodulosa* (Hymenoptera: Cynipidae) and those influenced by the inquiline *Periclistus pirata* (Hymenoptera: Cynipidae) on *Rosa blanda* (Rosaceae). *Can. J. Bot.* 76:365–381.

Brooks, Scott E., and Joseph D. Shorthouse. 1998b. Biology of the galler *Diplolepis rosaefolii* (Hymenoptera: Cynipidae), its associated component community, and host shift to the shrub rose Therese Bugnet. *Can. Entomol.* 130:357–366.

Brown, Leland R., and Clark O. Eads. 1965. A technical study of insects affecting the oak tree in southern California. *Univ. Calif. Agric. Exper. Station Bull.* 810:79–91.

Burdick, Donald J. 1967. Oviposition behavior and galls of *Andricus chrysolepidicola* (Ashmead) (Hymenoptera: Cynipidae). *Pan-Pac. Entomol.* 43 (3): 227–231.

Burnett, John A. 1974. A new cynipid wasp from California. *Pan-Pac. Entomol.* 50 (3): 298–302.

Caltagirone, L.E. 1964. Notes on the biology, parasites, and inquilines of *Pontania pacifica* (Hymenoptera: Tenthredinidae), a leaf-gall incitant on *Salix lasiolepis*. *Ann. Entomol. Soc. Am.* 57 (3): 279–291.

Chestnut, V.K. 1974. Plants used by the Indians of Mendocino County, California. *Contrib. U.S. Natl. Herbarium* 7:343–344.

Collier, Mary E.T., and Sylvia Barker Thalman. 1991. Interviews with Tom Smith and Maria Copa-Isabel Kelly's ethnographic notes on the Coast Miwok Indians of Marin and southern Sonoma Counties, California. *Miwok Archaeol. Preserve Marin Mapom Occ. Pap.* 6:100, 211.

Cranshaw, Whitney. 2004. *Garden insects of North America*. Princeton, NJ: Princeton University Press..

Dailey, D. Charles. 1969. Synonymy of *Dryocosmus attractans* (Kinsey) and *Callirhytis uvellae* (Weld)(Hymenoptera: Cynipidae). *Pan-Pac. Entomol.* 45 (2): 132–134.

Dailey, D. Charles. 1972. Contiguous areas of Arizona and Pacific Slope floras in northern Baja, California, Mexico. *Pan-Pac. Entomol.* 48 (1): 74–75.

Dailey, D. Charles. 1977. Elevation of *Loxaulus brunneus* variety *atrior* (Kinsey) to full species status (Hymenoptera: Cynipidae). *Pan-Pac. Entomol.* 53:145–146.

Dailey, D. Charles, and Linda Campbell. 1973. A new species of *Diplolepis* from California (Hymenoptera: Cynipidae). *Pan-Pac. Entomol.* 49 (2): 174–176.

Dailey, D. Charles, and A.S. Menke. 1980. Nomenclatorial notes on North American Cynipidae (Hymenoptera). *Pan-Pac. Entomol.* 56 (3): 170–174.

Dailey, D. Charles, and Christine M. Sprenger. 1973a. Unisexual generation of *Andricus atrimentus* (Hymenoptera: Cynipidae). *Pan-Pac. Entomol.* 49 (2): 171–173.

Dailey, D. Charles, and Christine M. Sprenger. 1973b. Synonymy of *Andricus gigas* and the bisexual generation of *Andricus crenatus* (Hymenoptera: Cynipidae). *Pan-Pac. Entomol.* 49 (3): 188–191.

Dailey, D. Charles, and Christine M. Sprenger. 1977. Three new gall-inducing *Callirhytis* Foerster from *Quercus cedrosensis* Muller (Hymenoptera: Cynipidae). *Pan-Pac. Entomol.* 53 (1): 43–46.

Dailey, D. Charles, and Christine M. Sprenger. 1983. Gall-inducing cynipid wasps from *Quercus dunnii* Kellogg (Hymenoptera). *Pan-Pac. Entomol.* 59 (1–4): 42–49.

Dailey, D. Charles, Tim Perry, and Christine M. Sprenger. 1974. Biology of three *Callirhytis* gall wasps from Pacific Slope *Erythrobalanus* oaks (Hymenoptera: Cynipidae). *Pan-Pac. Entomol.* 50 (1): 60–67.

Doutt, Richard L. 1960. Heterogony in *Andricus crystallinus* Bassett (Hymenoptera: Cynipidae). *Pan-Pac. Entomol.* 36 (4): 167–170.

Ehler, L.E. 1987. Ecology of *Rhopalomyia californica* Felt at Jasper Ridge (Diptera: Cecidomyiidae). *Pan-Pac. Entomol.* 63 (3): 237–241.

Eliason, Eileen A., and Daniel A. Potter. 2001. Spatial distribution and parasitism of leaf galls induced by *Callirhytis cornigera* (Hymenoptera: Cynipidae) on pin oak. *Environ. Entomol. Soc. Am.* 30 (2): 280–287.

Essig, E.O. 1926. *Insects of western North America.* New York: Macmillan Publishing.

Evans, D. 1967. The bisexual and agamic generations of *Besibicus mirabilis* (Hymenoptera: Cynipidae), and their associate insects. *Can. Entomol.* 99:187–196.

Evans, David. 1972. Alternate generations of gall cynipids (Hymenoptera: Cynipidae) on Garry oak. *Can. Entomol.* 104:805–818.

Evans, H. E. 1968. *Life on a little-known planet.* New York: E. P. Dutton.

Felt, Ephram Porter. 1965. *Plant galls and gall makers.* New York: Hafner Publishing.

Foote, Richard H., and F. L. Blanc. 1963. *The fruit flies or Tephritidae of California.* Bulletin of the California Insect Survey. Berkeley and Los Angeles: University of California Press.

Force, Don C. 1974. Ecology of insect host-parasitoid communities. *Science* 184:624–632.

Fronk, W. D., A. A. Beetle, and D. G. Fullerton. 1964. Dipterous galls on the *Artemisia tridentata* complex and insects associated with them. *Ann. Entomol. Soc. Am.* 57:575–577.

Furniss, Malcolm M., and William F. Barr. 1975. *Insects affecting important native shrubs of the northwestern United States.* U.S. Department of Agriculture Forest Service General Technical Report INT-19.

Furniss, R. L., and V. M. Carolin. 1977. *Western forest insects.* U.S. Department of Agriculture Forest Service Miscellaneous Publication 1339.

Gagné, Raymond J. 1973. A generic synopsis of the Nearctic Cecidomyiidae (Diptera: Cecidomyiidae: Cecidomyiinae). *Ann. Entomol. Soc. Am.* 66 (4): 857–889.

Gagné, Raymond J. 1975. The gall midges of ragweed, *Ambrosia,* with descriptions of two new species (Diptera: Cecidomyiidae). *Proc. Entomol. Soc. Wash.* 77 (1): 50–55.

Gagné, Raymond J. 1986. Revision of *Prodiplosis* (Diptera: Cecidomyiidae) with descriptions of three new species. *Ann. Entomol. Soc. Am.* 79 (1): 235–245.

Gagné, Raymond J. 1989. *The plant-feeding gall midges of North America.* Ithaca, NY: Cornell University Press.

Gagné, Raymond J. 2004. *A catalog of the Cecidomyiidae (Diptera) of the world.* Memoirs of the Entomological Society of Washington 25. Washington, DC: Entomological Society of Washington.

Gagné, Raymond J., and Paul E. Boldt. 1995. The gall midges (Diptera: Cecidomyiidae) of *Baccharis* spp. (Asteraceae) in the United States. *Proc. Entomol. Soc. Wash.* 97 (4): 767–778.

Gagné, Raymond J., and Bradford A. Hawkins. 1983. Biosystematics of the Lasiopterini (Diptera: Cecidomyiidae: Cecidomyiinae) associated with *Atriplex* spp. (Chenopodiaceae) in southern California. *Ann. Entomol. Soc. Am.* 76 (3): 379–383.

Gagné, Raymond J., and Donald R. Strong. 1993. A new species of *Dasineura* (Diptera: Cecidomyiidae) galling leaves of *Lupinus* spp. (Fabaceae) in California. *Proc. Entomol. Soc. Wash.* 95 (4): 541–546.

Gagné, Raymond J., and Gwendolyn L. Waring. 1990. The *Asphondylia* (Cecidomyiidae: Diptera) of creosote bush (*Larrea tridentata*) in North America. *Proc. Entomol. Soc. Wash.* 92 (4): 649–671.

Gambino, Parker. 1990. Mark-recapture studies on *Vespula pensylvanica* queens (Hymenoptera: Vespidae). *Pan-Pac. Entomol.* 66 (3): 227–231.

Gassmann, Andre, and Joseph D. Shorthouse. 1990. Structural damage and gall induction by *Pegomya curticornis* and *Pegomya euphorbiae* (Diptera: Anthomyiidae) within the stems of leafy spurge (*Euphorbia X Pseudovirgata*) (Euphorbiaceae). *Can. Entomol.* 122:429–439.

Goeden, Richard D. 1987. Life history of *Trupanea conjuncta* (Adams) on *Trixus californica* (Kellogg) in southern California (Diptera: Tephritidae). *Pan-Pac. Entomol.* 63 (3): 284–291.

Goeden, Richard D. 1988. Gall formation by the capitulum-infesting fruit fly, *Tephritis stigmata* (Diptera: Tephritidae). *Proc. Entomol. Soc. Wash.* 90 (1): 37–43.

Goeden, Richard D. 1990. Life history of *Eutreta diana* on *Artemisia tridentata* in southern California (Diptera: Tephritidae). *Pan-Pac. Entomol.* 66 (1): 24–32.

Goeden, Richard D. 2002a. Life history and description of immature stages of *Oxyna palpalis* (Coquillett) (Diptera: Tephritidae) on *Artemisia tridentata* (Nuttall) (Asteraceae) in southern California. *Proc. Entomol. Soc. Wash.* 104 (3): 537–553.

Goeden, Richard D. 2002b. Life history and description of immature stages of *Oxyna aterrima* (Doane) (Diptera: Tephritidae) on *Artemisia tridentata* (Nuttall) (Asteraceae) in southern California. *Proc. Entomol. Soc. Wash.* 104 (2): 510–526.

Goeden, Richard D. 2002c. Description of immature stages of *Tephritis stigmata* (Coquillett) (Diptera: Tephritidae). *Proc. Entomol. Soc. Wash.* 104 (2): 335–347.

Goeden Richard D., and David H. Headrick. 1991. Life history and descriptions of immature stages of *Tephtritis baccharis* (Coquillett) on *Baccharis salicifolia* (Ruiz and Pavon) Persoon in southern California (Diptera: Tephritidae). *Pan-Pac. Entomol.* 67 (2): 86–98.

Goeden, Richard D., and Allen L. Norrbom. 2001. Life history and description of adults and immature stages of *Procecidochares blanci,* new species (Diptera: Tephritidae) on *Isocoma acradenia* (E. Greene) (Asteraceae) in southern California. *Proc. Entomol. Soc. Wash.* 103 (3): 517–540.

Goeden, Richard D., and Jeffrey A. Teerink. 1996a. Life histories and descriptions of adults and immature stages of two cryptic species, *Aciurina ferruginea* and *A. michaeli,* new species (Diptera: Tephritidae) on *Chrysothamnus viscidiflorus* in southern California. *Proc. Entomol. Soc. Wash.* 98 (3): 415–438.

Goeden, Richard D., and Jeffrey A. Teerink. 1996b. Life history and descriptions of adults and immature stages of *Aciurina idahoensis* (Steyskal) (Diptera: Tephritidae) on *Chrysothamnus viscidiflorus* (Hooker) (Nuttall) in southern California. *Proc. Entomol. Soc. Wash.* 98 (4): 681–694.

Goeden, Richard, and Jeffrey A. Teerink. 1996c. Life history and descriptions of adults and immature stages of *Aciurina semilucida* (Diptera: Tephritidae) on *Chrysothamnus viscidiflorus* in southern California. *Proc. Entomol. Soc. Wash.* 98 (4): 752–766.

Goeden, Richard D., and Jeffrey A. Teerink. 1997a. Notes on the life histories and descriptions of adults and immature stages of *Procecidochares kristineae* and *P. lisae* new species (Diptera: Tephritidae) on *Ambrosia* spp. in southern California. *Proc. Entomol. Soc. Wash.* 99 (1): 67–88.

Goeden, Richard D., and Jeffrey A. Teerink. 1997b. Life history and description of immature stages of *Procecidochares anthracina* (Doane) (Diptera: Tephritidae) on *Solidago californica* (Nuttall) in southern California. *Proc. Entomol. Soc. Wash.* 99 (1): 180–193.

Goeden, Richard D., and Jeffrey A. Teerink. 1997c. Life history and description of immature stages of *Trupanea signata* (Foote) (Diptera: Tephritidae) on *Gnaphalium luteo-album* L. in southern California. *Proc. Entomol. Soc. Wash.* 99 (4): 748–755.

Goeden, Richard D., David H. Headrick, and Jeffrey Teerink. 1995. Life history description of immature stages of *Valentibulla californica* (Coquillett) (Diptera: Tephritidae) on *Chrysothamnus nauseosus* (Pallas) (Britton) in southern California. *Proc. Entomol. Soc. Wash.* 97 (3): 548–560.

Goodrich, Jennie, Claudia Lawson, and Vana Parrish Lawson. 1980.

Kashaya Pomo plants. American Indian Monograph Series, vol. 2, no. 79. Los Angeles: University of California.

Gordh, Gordon, and Bradford A. Hawkins. 1982. *Tetrastichus cecidobroter* (Hymenoptera: Eulophidae), a new phytophagus species developing within galls of *Asphondylia* (Diptera: Cecidomyiidae) on *Atriplex* (Chenopodiaceae) in southern California. *Proc. Entomol. Soc. Wash.* 84 (3): 426–429.

Grigarick, A.A., and W.H. Lange. 1968. Seasonal development and emergence of two species of gall-forming aphids, *Pemphigus bursarius* and *P. nortoni,* associated with poplar trees in California. *Ann. Entomol. Soc. Am.* 61 (2): 509–514.

Harper, A.M. 1959. Gall aphids on poplar in Alberta: Descriptions of galls and distribution of aphids. *Can. Entomol.* 91 (8): 489–496.

Harper, A.M. 1966. Three additional poplar gall aphids from southern Alberta. *Can. Entomol.* 98:1212–1214.

Harris, M.O., J.J. Stuart, M. Mohan, S. Nair, R.J. Lamb, and O. Rohfritsch. 2003. Grasses and gall midges: Plant defense and insect adaptation. *Annu. Rev. Entomol.* 48:549–577.

Hartman, Hollister. 1984. Ecology of gall-forming Lepidoptera on *Tetradymia. Hilgardia* 52 (3): 1–39.

Harville, John P. 1955. Ecology and population dynamics of the California Oak Moth, *Phryganidia californica* Packard (Lepidoptera: Dioptidae). *Microentomology* 20 (4): 83–166.

Hawkins, Bradford A., and Richard D. Goeden. 1982. Biology of gallforming *Tetrastichus* (Hymenoptera: Eulophidae) associated with gall midges on saltbush in southern California. *Entomol. Soc. Am.* 75 (4): 444–447.

Hawkins, Bradford A., and Richard D. Goeden. 1984. Organization of a parasitoid community associated with a complex of galls on *Atriplex* spp. in southern California. *Ecol. Entomol.* 9:271–292.

Hawkins, Bradford A., and Thomas R. Unruh. 1988. Protein and water levels in *Asphondylia atriplicis* (Diptera: Cecidomyiidae) galls. *Southwestern Nat.* 33 (1): 114–117.

Hawkins, Bradford A., Richard D. Goeden, and Raymond J. Gagné. 1986. Ecology and taxonomy of the *Asphondylia* spp. (Diptera: Cecidomyiidae) forming galls on *Atriplex* spp. (Chenopodiaceae) in southern California. *Entomography* 4:55–107.

Haws, Austin, Alan H. Roe, and David L. Nelson. 1988. *Index to information on insects associated with western wildland shrubs.* U.S. Department of Agriculture Forest Service General Technical Report INT-248.

Headrick, David H., and Richard D. Goeden. 1993. Life history and description of immature stages of *Aciurina thoracica* (Diptera: Tephritidae) on *Baccharis sarothroides* in southern California. *Entomol. Soc. Am.* 86 (1): 68–79.

Headrick, David H., and Richard D. Goeden. 1997. Gall midge forms galls on fruit fly galls (Diptera: Cecidomyiidae, Tephritidae). *Proc. Entomol. Soc. Wash.* 99 (3): 487–489.

Headrick, David H., and Richard D. Goeden. 1998. The biology of non-frugivorous tephritid fruit flies. *Annu. Rev. Entomol.* 43: 217–241.

Headrick, David H., Richard D. Goeden, and Jeffrey A. Teerink. 1997. Taxonomy of *Aciurina trixa* (Curran) (Diptera: Tephritidae) and its life history on *Chrysothamnus nauseosus* (Pallas) (Britton) in southern California; with notes on *A. bigeloviae* (Cockerell). *Proc. Entomol. Soc. Wash.* 99 (3): 415–428.

Hepting, George H. 1971. *Diseases of forest and shade trees of the United States.* U.S. Department of Agriculture Forest Service Agricultural Handbook 386.

Heydon, Steven L. 1994. Taxonomic changes in Nearctic Pteromalidae. II: New synonymy and four new genera (Hymenoptera: Chalcidoidea). *Proc. Entomol. Soc. Wash.* 96 (2): 323–338.

Hickman, James C., ed. 1993. *The Jepson manual: Higher plants of California.* Berkeley and Los Angeles: University of California Press.

Hildebrand, D. C., and M. N. Schroth. 1967. A new species of *Erwinia* causing the drippy nut disease of live oaks. *Phytopathology* 57 (3): 250–253.

Hufbauer, Ruth A. 2004. Observations of sagebrush gall morphology and emergence of *Rhopalomyia pomum* (Diptera: Cecidomyiidae) and its parasitoids. *Western N. Am. Nat.* 64 (3): 324–330.

Hutchins, Ross E. 1969. *Galls and gall insects.* New York: Dodd, Mead and Co.

Johnson, Warren T., and Howard H. Lyon. 1991. *Insects that feed on trees and shrubs.* Ithaca, NY: Cornell University Press.

Jones, Robert G., Raymond J. Gagné, and William F. Barr. 1983. Biology and taxonomy of the *Rhopalomyia* gall midges (Diptera: Cecidomyiidae) of *Artemisia tridentata* (Compositae) in Idaho. *Contrib. Am. Entomol. Inst.* 21 (1): 1–76.

Keifer, H. H. 1952. *The eriophyid mites of California.* Bulletin of the California Insect Survey, vol. 2, no. 1. Berkeley and Los Angeles: University of California Press.

Keifer, Hartford H., Edward W. Baker, Tokuwo Kono, Mercedes Delfinado, and William E. Styer. 1982. *An illustrated guide to plant abnormalities caused by eriophyid mites in North America.* U.S. Department of Agriculture Handbook 573.

Kinsey, Alfred C. 1922. Studies of some new and described Cynipidae (Hymenoptera). *Indiana Univ. Studies* 53:3–141.

Lange, W. H. 1965. Biosystematics of American *Pemphigus* (Homoptera: Aphidoidea). In *XIIth International Congress of Entomology,* 102–104.

Larew, Hiram, and Joseph Capizzi. 1983. *Common insect and mite galls of the Pacific Northwest.* Corvallis: Oregon State University Press.

Leech, Hugh B. 1948. Elm gall aphid eaten by Evening Grosbeak. *Entomol. Soc. Brit. Columbia* 44:34.

Lyon, Robert J. 1959. An alternating, sexual generation in the gall wasp *Callirhytis pomiformis* (ASHM) (Hymenoptera: Cynipidae). *Bull. Southern Calif. Acad. Sci.* 58 (1): 33–37.

Lyon, Robert J. 1963. The alternate generation of *Heteroecus pacificus* (Ashmead) (Hymenoptera: Cynipidae). *Proc. Entomol. Soc. Wash.* 65 (3): 250–254.

Lyon, Robert J. 1964. The alternate generation of *Callirhytis agrifoliae* (Ashmead) (Hymenoptera: Cynipidae). *Proc. Entomol. Soc. Wash.* 66 (3): 193–196.

Lyon, Robert J. 1969. An alternate generation of *Callirhytis quercussuttoni* (Bassett) (Hymenoptera: Cynipidae). *Proc. Entomol. Soc. Wash.* 71 (1): 61–65.

Lyon, Robert J. 1970. Heterogony in *Callirhytis serricornis* (Kinsey) (Hymenoptera: Cynipidae). *Proc. Entomol. Soc. Wash.* 72 (2): 176–178.

Lyon, Robert J. 1993. Synonymy of two genera of cynipid wasps and description of a new genus (Hymenoptera: Cynipidae). *Pan-Pac. Entomol.* 69 (2): 133–140.

Lyon, Robert J. 1996. New cynipid wasps from the southwestern United States (Hymenoptera: Cynipidae). *Pan-Pac. Entomol.* 72 (4): 181–192.

MacKay, Pam. 2003. *Mojave Desert wildflowers.* Falcon Guide. Guilford, CT: Globe Pequot Press.

Mani, M. S. 1964. *Ecology of plant galls.* The Hague: D. W. Junk, Publishers.

McGavin, George C. 2002. *Insects, spiders, and other terrestrial arthropods.* Smithsonian Handbooks. Washington, DC: American Museum of Natural History.

McKell, Cyrus M., James R. Blaisdell, and Joe R. Gordon. 1972. *Wildland shrubs: Their biology and utilization.* U.S. Department of Agriculture Forest Service General Technical Report INT-1.

Melika, George, and Warren G. Abrahamson. 2002. Review of the world genera of oak cynipid wasps (Hymenoptera: Cynipidae: Cynipini). In *International symposium: Parasitic wasps: Evolution, systematics, biodiversity and biological control,* ed. G. Melika and C. Thuroczy, 150–190. Köszeg, Hungary: Talajvédelmi Szolgálat Rovar Parasitology Laboratory. http://ecol1.bio.u-szeged.hu/~rovpar/symposium/melikaabrahamson.pdf (accessed May 2006).

Meyer, Jean. 1987. *Plant galls and gall inducers.* Berlin: Gebruder Borntraeger.

Miller, D. G., III. 1998. Life history, ecology, and communal gall occupation in the manzanita leaf-gall aphid, *Tamalia coweni* (Cockerell) (Homoptera: Aphididae). *J. Nat. Hist.* 32:351–366.

Miller, Donald G., III. 2004. The ecology of inquilinism in communally parasitic *Tamalia* aphids (Hemiptera: Aphididae). *Ann. Entomol. Soc. Am.* 97 (6): 1233–1241.

Miller, Donald G., and Michael J. Sharkey. 2000. An inquiline species of *Tamilia* co-occurring with *Tamalia coweni* (Homoptera: Aphididae). *Pan-Pac. Entomol.* 75 (2): 77–86.

Miller, William E. 2000. A comparative taxonomic–natural history study of eight Nearctic species of *Gnorimoschema* that induce stem galls on Asteraceae, including descriptions of three new species (Lepidoptera: Gelechiidae). Thomas Say Publications in Entomology. Lanham, MD: Entomological Society of America.

Miller, William E. 2005. Gall-inducing lepidoptera. In *Biology, ecology, and evolution of gall-inducing arthropods,* ed. Anantanarayanan Raman, Carl W. Schaefer, and Toni M. Withers, 431–465. Enfield, NH: Science Publishers.

Mitchell, R. G., and J. K. Maksymov. 1977. Observations of predation on spruce gall aphids within the gall. *Entomophaga* 22 (2): 179–186.

Mix, A. J. 1949. A monograph of the genus *Taphrina. Sci. Bull. Univ. Kansas* 33 (1): 3–167.

Moran, Nancy A. 1992. Quantum leapers. *Natural History,* April, 34–39.

Munz, Phillip A. 1963. *A flora of California.* Berkeley and Los Angeles: University of California Press.

Nyman, Tommi. 2000. Phyolgeny and ecological evolution of gall-inducing sawflies (Hymenoptera: Tenthredinidae). Ph.D. dissertation in biology. University of Joensuu, Finland.

Oldfield, G.N., R.F. Hobza, and N.S. Wilson. 1970. Discovery and characterization of spermatophores in the eriophyoidea (Acari). *Ann. Entomol. Soc. Am.* 63 (2): 520–526.

Palmer, Miriam A. 1952. *Aphids of the Rocky Mountains region, subfamily Eriosomatinae,* 359–365. Thomas Say Foundation.

Phillips, Steven J., and Patricia Wentworth Comus, eds. 2000. *A natural history of the Sonoran Desert.* Berkeley and Los Angeles: Arizona Desert Museum and University of California Press.

Pojar, Jim, and Andy MacKinnon, eds. 1994. *Plants of the Pacific Northwest coast.* Edmonton, Canada: British Columbia Ministry of Forests and Lone Pine Publishing.

Povolny, D. 2003. Description of twenty-five new Nearctic species of the genus *Gnorimoschema* Busck, 1900 (Lepidoptera: Gelechiidae). *SHILAP Revta. Lepid.* 31 (124): 285–315.

Powell, Jerry A. 1975. Biological records and descriptions of some little known *Epiblema* in the southwestern United States (Lepidoptera: Tortricidae). *Pan-Pac. Entomol.* 51 (2): 99–112.

Powell, Jerry A., and Charles L. Hogue. 1979. *California insects.* Berkeley and Los Angeles: University of California Press.

Powell, Jerry A., and Dalibor Povolny. 2001. Gnorimoschemine moths of the coastal dune and scrub habitats in California (Lepidoptera: Gelechiidae). *Holarctic Lepid.* 8 (1): 1–53.

Pritchard, Earl A. 1953. *The gall midges of California.* Bulletin of the California Insect Survey. Berkeley and Los Angeles: University of California Press.

Rosenthal, Sarah Suzanne. 1968. Biology and host relations of some Cynipidae forming galls on *Quercus.* Doctoral dissertation. University of California, Berkeley.

Rosenthal, S.S., and C.S. Koehler. 1971a. Heterogony in some gall-forming Cynipidae (Hymenoptera) with notes on the biology of *Neuroteras saltatorius. Ann. Entomol. Soc. Am.* 64 (3): 565–570.

Rosenthal, S.S., and C.S. Koehler. 1971b. Intertree distributions of some cynipid (Hymenoptera) galls on *Quercus lobata. Ann. Entomol. Soc. Wash.* 64 (3): 571–574.

Russo, Ron. 1975. Gall wasp nurseries. *Pac. Discovery* 28 (6): 25–31.

Russo, Ronald A. 1979. *Plant galls of the California region.* Pacific Grove, CA: Boxwood Press.

Russo, Ron. 1981. Oak galls. *Monterey Life,* January, 80–81.

Russo, Ron. 1983. Galls: Surreal ornaments on blue oaks. *Fremontia* 11 (3): 19–22.

Russo, Ron. 1990. Blue oak: A gall wasp nursery. *Fremontia* 18 (3): 68–71.

Russo, Ron. 1991. *Gall wasps and oaks: Oaks of California.* Los Olivos, CA: Cachuma Press.

St. John, Mark G., and Joseph D. Shorthouse. 2000. Allocation patterns of organic nitrogen and mineral nutrients within stem galls of *Diplolepis spinosa* and *Diplolepis triforma* (Hymenoptera: Cynipidae) on wild roses (Rosaceae). *Can. Entomol.* 132:635–648.

Sanver, Dilek, and Bradford A. Hawkins. 2000. Galls as habitats: The inquiline communities of insect galls. *Basic Appl. Ecol.* 1:3–11.

Scharpf, Robert F. 1993. *Diseases of Pacific Coast conifers.* U.S. Department of Agriculture Forest Service Handbook 521.

Scher, Stanley, and Gretchen Wilson. 1996. Tumorigenesis in coast redwoods: Analysis of tumor occurrence and severity in northern California. In *Proceedings of the conference on coast redwood forest ecology and management,* 99–101. Arcata, CA: Humboldt State University.http://web.archive.org/web/20040624200714/cnr.ber keley.edu/~jleblanc/WWW/Redwood/rdwd-TUMORIGE.html (accessed May 2006).

Schick, Katherine N., and Donald L. Dahlsten. 2003. Gallmaking and insects. In *Encyclopedia of insects,* 464–466. New York: Academic Press.

Shorthouse, J.D. 1973a. *Common insect galls of Saskatchewan.* New York: Bluejay Books.

Shorthouse, J.D. 1973b. The insect community associated with rose galls of *Diplolepis polita* (Cynipidae, Hymenoptera). *Quaestion. Entomol.* 9:55–98.

Shorthouse, J.D. 1993. Adaptations of gall wasps of the genus *Diplolepis* (Hymenoptera: Cynipidae) and the role of gall anatomy in cynipid systematics. *Mem. Entomol. Soc. Can.* 165:139–163.

Shorthouse, Joseph D. 2001. Galls induced by cynipid wasps of the genus *Diplolepis* (Cynipidae, Hymenoptera) on cultivated shrub roses in Canada. *ISHS Acta Hort.* 547:83–92.

Shorthouse, Joseph D., and Scott E. Brooks. 1998. Biology of the galler *Diplolepis rosaefolii* (Hymenoptera: Cynipidae), its associated component community, and host shift to the shrub rose Therese Bugnet. *Can. Entomol.* 130:357–366.

Shorthouse, Joseph D., and Andre Gassmann. 1994. Gall induction by *Pegomya curticornis* (Stein) (Diptera: Anthomyiidae) within the roots of spurges *Euphorbia virgata* Waldst. and Kit. and *E. esula* L. (Euphorbiaceae). *Can. Entomol.* 126:193–197.

Shorthouse, Joseph D., and Odette Rohfritsch, eds. 1992. *Biology of insect-induced galls.* Oxford, UK: Oxford University Press.

Silverman, J., and R. D. Goeden. 1980. Life history of a fruit fly, *Procecidocharaes* sp., on the ragweed *Ambrosia dumosa* in southern California (Diptera: Tephritidae). *Pan-Pac. Entomol.* 56 (4): 283–288.

Sinclair, Wayne A., Howard H. Lyon, and Warren T. Johnson. 1987. *Diseases of trees and shrubs.* Ithaca, NY: Cornell University Press.

Smith, Edward Laidlaw. 1968. Biosystematics and morphology of symphyta. I. Stem-galling *Euura* of the California region, and a new female genitalic nomenclature. *Ann. Entomol. Soc. Am.* 61 (6): 1389–1407.

Smith, Edward Laidlaw. 1970. Biosystematics and morphology of symphyta. II. Biology of gall-making nematine sawflies in the California region. *Ann. Entomol. Soc. Am.* 63 (1): 36–51.

Spencer, Kenneth A., and Bradford A. Hawkins. 1984. An interesting gall-forming *Ophiomyia* species (Diptera: Agromyzidae) on *Atriplex* (Chenopodiaceae) in southern California. *Proc. Entomol. Soc. Wash.* 86 (3): 664–668.

Stone, Graham N., Karsten Schonrogge, Rachel J. Atkinson, David Bellido, and Juli Pujade-Villar. 2002. The population biology of oak gall wasps (Hymenoptera: Cynipidae). *Annu. Rev. Entomol.* 47:633–668.

Sullivan, Daniel J. 1987. Insect hyperparasitism. *Annu. Rev. Entomol.* 32:49–70.

Swanton, E. W. 1912. *British plant galls.* London: Methuen and Co.

Tauber, Maurice J., and Catherine A. Tauber. 1968. Biology of the gall-former *Procecidochares stonei* on a composite. *Ann. Entomol. Soc. Am.* 61 (2): 553–554.

Tilden, James W. 1951. The insect asssociates of *Baccharis pilularis.* *Microentomol. Stanford Univ. Press* 16 (1): 149–188.

Walls, Lee, and Benjamin A. Zamora. 2001. Nitrogen-fixing nodule characterization and morphology of four species in the northern intermountain region. In *U.S. Department of Agriculture Forest Service Proceedings* RMRS-P-21, 295–301.

Wangberg, James K. 1975. Biology of the thimbleberry gallmaker *Diastrophus kincaidii* (Hymenoptera: Cynipidae). *Pan-Pac. Entomol.* 51 (1): 39–48.

Wangberg, James K. 1980. Comparative biology of gall-formers in the genus *Procecidochares* (Diptera: Tephritidae) on rabbitbrush in Idaho. *J. Kansas Entomol. Soc.* 53 (2): 401–420.

Wangberg, James K. 1981a. Gall-forming habits of *Aciurina* species (Diptera: Tephritidae) on rabbitbrush (Compositae: *Chrysothamnus spp.*) in Idaho. *J. Kansas Entomol. Soc.* 54 (4): 711–732.

Wangberg, James K. 1981b. Observation on the bionomics of *Rhopalomyia utahensis* Felt (Diptera: Cecidomyiidae), its gall and insect associates. *Cecid. Int.* 2 (1): 17–24.

Waring, Gwen. 1986. Creosote bush: The ultimate desert survivor. *Agave* 2 (1): 3–15.

Waring, Gwendolyn L., and Peter W. Price. 1989. Parasitoid pressure and the radiation of a gallforming group (Cecidomyiidae: *Asphondylia* spp.) on creosote bush *(Larrea tridentata).* *Oecologia* 79:293–299.

Waring, Gwendolyn L., and Peter W. Price. 1990. Plant water stress and gall formation (Cecidomyiidae: *Asphondylia* spp.) on creosote bush. *Ecol. Entomol.* 15:87–95.

Washburn, Jan O., and Howard V. Cornell. 1979. Chalcid parasitoid attack on a gall wasp population, *Acraspis hirta* (Hymenoptera: Cynipidae) on *Quercus prinus* (Fagaceae). *Can. Entomol.* 111: 391–400.

Weis, Arthur E., Rod Walton, and Cathryn L. Crego. 1988. Reactive plant tissue sites and the population biology of gall makers. *Annu. Rev. Entomol.* 33:476–486.

Weld, Lewis H. 1952a. New American cynipid wasps from galls. *U.S. Natl. Mus. Proc.* 102:327–328.

Weld, Lewis H. 1952b. *Cynipoidae (Hymenoptera), 1905–1950.* Privately published.

Weld, Lewis H. 1957. *Cynipid galls of the Pacific Slope.* Privately published.

Weld, Lewis H. 1960. *Cynipid galls of the Southwest.* Privately published.

Westcott, Cynthia. 1971. *Plant disease handbook.* New York: Van Nostrand Reinhold Co.

Williams, Jason B., Joseph D. Shorthouse, and Richard E. Lee Jr. 2002. Extreme resistance to desiccation and microclimate-related differences in cold-hardiness of gall wasps (Hymenoptera: Cynipidae) overwintering on roses in southern Canada. *J. Exp. Biol.* 205: 2115–2124.

Wood, David L., Thomas W. Koerber, Robert F. Scharpf, and Andrew J. Storer. 2003. *Pests of the native California conifers.* Berkeley and Los Angeles: University of California Press.

Wool, David. 2004. Galling aphids: Specialization, biological complexity, and variation. *Annu. Rev. Entomol.* 49:175–192.

Young, James A., and Charlie D. Clements. 2002. *Purshia: The wild and bitter roses.* Reno: University of Nevada Press.

Zigmond, Maurice L. 1981. *Kawaiisu ethnobotany.* Salt Lake City: University of Utah Press.

INDEX

Page references in **boldface type** refer to the main discussion of the topic.

ABOUT THE AUTHOR

Ron worked as a naturalist for the East Bay Regional Park District, Alameda and Contra Costa Counties, California, for 37 years. He retired as the chief naturalist in 2003 after having been in that position for 17 years. In 1989, he received the distinguished Fellow Award from the National Association for Interpretation, an international organization for professional naturalists and historic interpreters. Ron's passion for learning and nature has led him to specialize in mushrooms, nudibranchs and  other marine invertebrates, sharks, whales, wildflower pollination, chaparral, plant galls, and mammals. Ron has been guiding natural history trips in southeastern Alaska for 15 years. He is the author of six field guides including *Plant Galls of the California Region*, *Pacific Intertidal Life*, *Pacific Coast Fish*, *Pacific Coast Mammals*, *Mountain States Mammals*, and *Hawaiian Reefs*. In addition, he has published numerous articles and papers in natural history magazines and technical journals.

Series Design:	Barbara Jellow
Design Enhancements:	Beth Hansen
Design Development:	Jane Tenenbaum
Composition:	Jane Tenenbaum
Indexer:	Jeanne Moody
Text:	9/10.5 Minion
Display:	Franklin Gothic Book and Demi
Printer and binder:	Golden Cup Printing Company Limited

Introduction to California Desert Wildflowers, Revised Edition, by Philip A. Munz, edited by Diane L. Renshaw and Phyllis M. Faber

Introduction to California Plant Life, Revised Edition, by Robert Ornduff, Phyllis M. Faber, and Todd Keeler-Wolf

Introduction to California Chaparral, by Ronald D. Quinn and Sterling C. Keeley, with line drawings by Marianne Wallace

Introduction to the Plant Life of Southern California: Coast to Foothills, by Philip W. Rundel and Robert Gustafson

Introduction to Horned Lizards of North America, by Wade C. Sherbrooke

Introduction to the California Condor, by Noel F. R. Snyder and Helen A. Snyder

Regional Guides

Sierra Nevada Natural History, Revised Edition, by Tracy I. Storer, Robert L. Usinger, and David Lukas